The Open University

Mathematics Foundation Course Unit 31

DIFFERENTIAL EQUATIONS II

Prepared by the Mathematics Foundation Course Team

Correspondence Text 31

The Open University Press

Open University courses provide a method of study for independent learners through an integrated teaching system including textual material, radio and television programmes and short residential courses. This text is one of a series that make up the correspondence element of the Mathematics Foundation Course.

The Open University's courses represent a new system of university level education. Much of the teaching material is still in a developmental stage. Courses and course materials are, therefore, kept continually under revision. It is intended to issue regular up-dating notes as and when the need arises, and new editions will be brought out when necessary.

Further information on Open University courses may be obtained from The Admissions Office, The Open University, P.O. Box 48, Bletchley, Buckinghamshire.

The Open University Press
Walton Hall, Bletchley, Bucks

First Published 1971

Printed in Great Britain by
J W Arrowsmith Ltd, Bristol 3

SBN: 335 01030 X

Contents

Objectives

After working through this unit you should be able to:

(i) explain the meanings of the terms: complementary function, constant coefficients, equation of motion, forced vibration, free vibration, linear differential operator, order of a linear differential operator, frequency, period, mass, force, resonance;

(ii) formulate the differential equation of motion for a mass-spring system in which the mass moves in a straight line;

(iii) determine whether a given differential operator is linear or non-linear;

(iv) find particular solutions of a linear differential equation with constant coefficients;

(v) understand the statement of the theorem given on page 18 and apply the theorem in specific instances;

(vi) write down general solutions of linear differential equations, given enough particular solutions to span the solution space;

(vii) determine the values of arbitrary constants from initial conditions;

(viii) determine the period of a given function;

(ix) solve equations of the form

$$Lf(t) = k \cos pt \quad \text{or} \quad k \sin pt,$$

where

$$Lf = O$$

is a linear differential equation having enough particular solutions to span the solution space, and cos pt (or sin pt) $\notin$ the kernel of L.

Note

Before working through this correspondence text, make sure you have read the general introduction to the mathematics course in the Study Guide, as this explains the philosophy underlying the whole course. You should also be familiar with the section which explains how a text is constructed and the meanings attached to the stars and other symbols in the margin, as this will help you to find your way through the text.

Structural Diagram

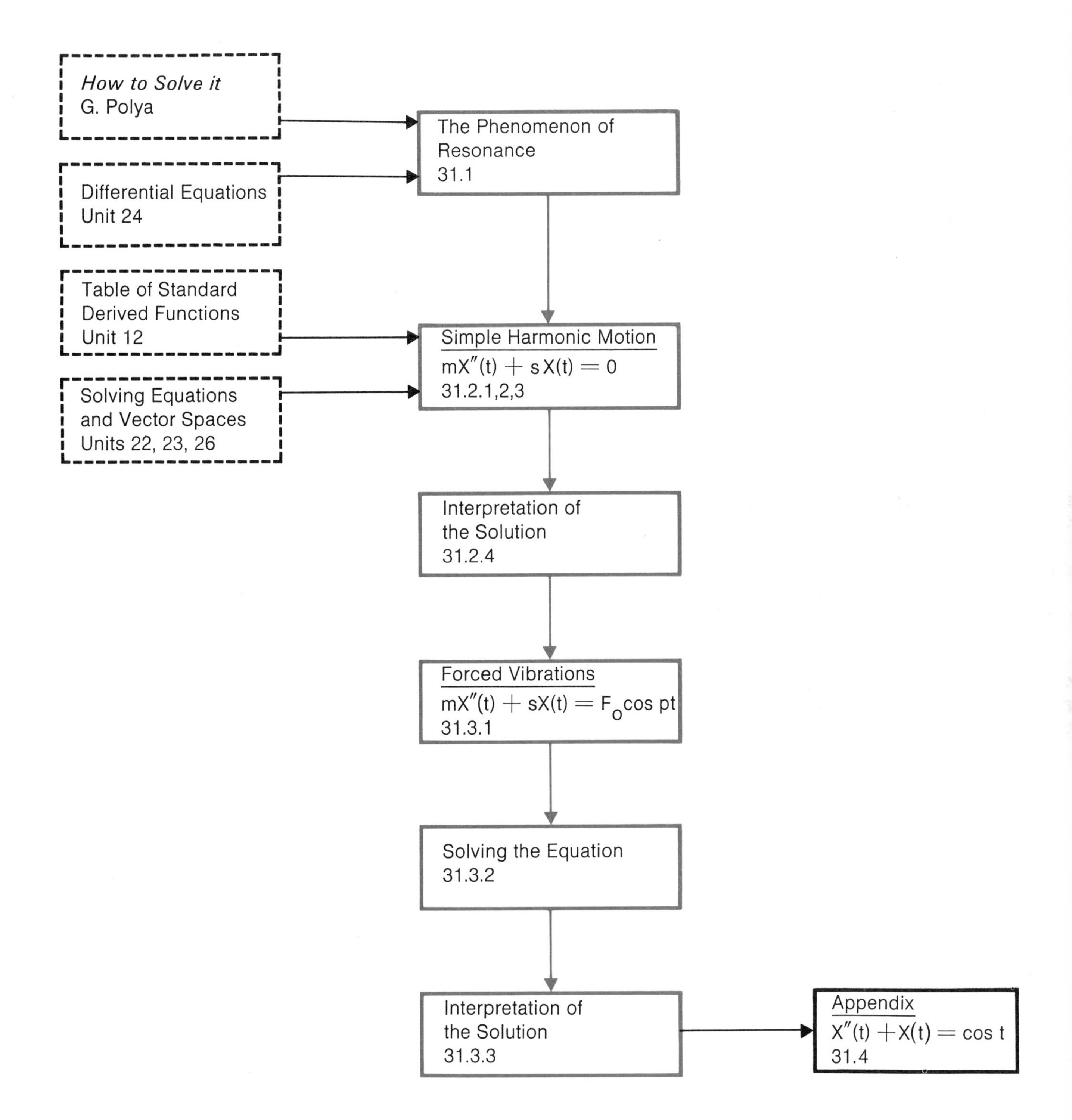

Glossary

Page

Terms which are defined in this glossary are printed in CAPITALS.

AMPLITUDE	The AMPLITUDE of a VIBRATION is the distance between the extreme position of the vibrating PARTICLE and the EQUILIBRIUM POSITION.	1
ARBITRARY CONSTANTS	ARBITRARY CONSTANTS are numbers used to label the various solutions of a DIFFERENTIAL EQUATION. For example, the constants A and B in the formula $X: t \longmapsto A \cos t + B \sin t$ for the general solution of $X'' + X = O$.	24
COEFFICIENTS	See LINEAR DIFFERENTIAL OPERATOR.	
COMPLEMENTARY FUNCTION	The COMPLEMENTARY FUNCTION of a LINEAR DIFFERENTIAL EQUATION $Lf = g$ is an element of the kernel of L.	29
DISPLACEMENT	The DISPLACEMENT of a PARTICLE in space is the directed distance from some fixed origin to the particle.	4
EQUATION OF MOTION	The EQUATION OF MOTION of a PARTICLE is the DIFFERENTIAL EQUATION obtained by applying NEWTON'S SECOND LAW to the particle.	9
EQUILIBRIUM POSITION	The EQUILIBRIUM POSITION of a PARTICLE is the position at which the particle can remain at rest if undisturbed.	4
EXTERNAL FORCE	An EXTERNAL FORCE on a PARTICLE is a term in the expression for the FORCE which is independent of the particle's position.	6
FORCE	Physically, FORCE is a primitive concept of mechanics and can be defined only with reference to experience. Mathematically, FORCE can be partially defined through the way in which it appears in NEWTON'S LAWS OF MOTION.	6
FORCED VIBRATION	A FORCED VIBRATION of a PARTICLE is a VIBRATION for which the expression for the FORCE includes an EXTERNAL FORCE.	6
FREE VIBRATION	A FREE (or NATURAL) VIBRATION of a PARTICLE is a VIBRATION for which the FORCE is completely determined by the position of the particle.	6
FREQUENCY	The FREQUENCY of a SIMPLE HARMONIC MOTION is the reciprocal of the PERIOD.	25
LAW OF MOTION	See NEWTON'S SECOND LAW.	
LINEAR DIFFERENTIAL EQUATION	A LINEAR DIFFERENTIAL EQUATION is a DIFFERENTIAL EQUATION of the form $Lf = g$, where L is a LINEAR DIFFERENTIAL OPERATOR.	24

		Page
LINEAR DIFFERENTIAL OPERATOR	A LINEAR DIFFERENTIAL OPERATOR is an operator of the form $$L: f \longmapsto k_0 f + \sum_{r=1}^{n} k_r D^r f$$ with domain a suitable set of real functions, where the COEFFICIENTS $k_0, k_1, \ldots, k_n$ are real functions.	17
MASS	The MASS of a PARTICLE is the ratio of the FORCE acting on the particle to the acceleration of the particle.	6
NATURAL VIBRATION	See FREE VIBRATION.	
NEWTON'S SECOND LAW	NEWTON'S SECOND LAW is a statement of the empirical fact that the acceleration of a PARTICLE is proportional to the FORCE acting on the particle.	6
ORDER	The ORDER of a LINEAR DIFFERENTIAL OPERATOR is the order of the highest-order derived function appearing in it with non-zero coefficient.	17
OSCILLATION	See VIBRATION.	
PARTICLE	A PARTICLE is a body which can be represented approximately by a movable mathematical point.	4
PERIOD	A function f with domain R has PERIOD h if $$f(x+h) = f(x)$$ for all $x \in R$, where $\|h\|$ is the smallest number with this property.	24
RESONANCE	RESONANCE is the tendency of the VIBRATIONS to reach very large AMPLITUDES when the EXTERNAL FORCE OSCILLATES with the PERIOD of the FREE VIBRATIONS of the system.	2
SIMPLE HARMONIC MOTION	SIMPLE HARMONIC MOTION is the motion of a PARTICLE whose EQUATION OF MOTION is $$mX'' + sX = O.$$	10
VIBRATION	A VIBRATION (or OSCILLATION) of a PARTICLE is a repetitive to-and-fro motion.	1

Notation

The symbols are presented in the order in which they appear in the text.

Symbol	Meaning	Page
R_0^+	The set of positive real numbers and zero.	5
m	The mass of a particle.	6
s	The stiffness of a spring.	7
O	The zero function: $O: x \longmapsto 0$ with domain a subset of R.	10
D	The differentiation operator.	11
ω	$\sqrt{\frac{s}{m}}$.	12
L	An operator.	14
k_i	A real function.	17
A	A linear transformation (morphism) of one vector space to another.	27
$\underline{x}$	An element of a vector space.	16, 27
$\underline{0}$	The zero element of a vector space.	16, 27

Bibliography

There are many books on mechanics and differential equations, all of them going much further than we have done in this unit. The following have the advantage of being useful for other parts of this or other courses as well.

T. M. Apostol, *Calculus*, Vol. I (Blaisdell, 1967).
This book is also recommended for some other units. Section 8 covers the mathematics of free vibrations. (If you have the 1961 edition, the relevant pages are 237–240.)

D. L. Kreider, R. G. Kuller, D. R. Ostberg and F. W. Perkins, *An Introduction to Linear Analysis* (Addison-Wesley, 1966).
This is one of the set books for the M201 linear mathematics course. It stresses the vector space aspect of linear differential equations. Linear differential operators and equations are considered on pages 86–95. The theorem about the dimension of the solution space is proved on pages 106–108 (as a consequence of another theorem about differential equations which is *not* proved there).

R. C. Smith and P. Smith, *Mechanics* (John Wiley, 1968).
This is the set book for the MST 282 mechanics and applied calculus course. The differential equations discussed in this unit are considered on pages 101–109 of the book, Newton's laws of motion on pages xv–xvii and 49–54, the behaviour of springs on pages 144–147, and the theory of vibrations on pages 172–184.

R. L. E. Schwarzenberger, *Elementary Differential Equations* (Chapman and Hall, 1969).
This little book uses a notation and approach similar to the one used in this course. Chapter 5 discusses linear differential equations.

31.1 THE PHENOMENON OF RESONANCE

31.1

31.1.0 Introduction

31.1.0

Introduction
* *

In this unit we shall draw together several strands of argument that have arisen at various stages of the course, and apply them in the discussion of a familiar physical phenomenon — *resonance*. The word *resonance* refers, in the first instance, to sound: it describes, for example, the acoustic property of bathrooms that makes them flatter our singing voices. However, we shall use the word in a more general sense, to refer not only to sound but also to any kind of vibration, such as the vibrations of an electric motor, or of a car being driven over a bumpy road, or even the electrical vibrations produced in a T.V. or radio set by the electromagnetic waves from the transmitter. We shall describe enough of the physical laws relevant to our problems for the text to be intelligible without prior knowledge of mechanics.

You can easily demonstrate resonance for yourself. Hang any small object, for example, a key, from a string or chain a foot or two in length.

Hold the top end of the string or chain in your hand and move it regularly back and forth in a horizontal line a few inches in length. If you do this slowly, the object will follow your hand; if you do it rapidly, the object will hardly move; but if you do it at just the right frequency the object will swing back and forth with a much larger amplitude than that of your hand, and you will find that an almost imperceptible hand movement is enough to keep the object swinging violently.

A more dangerous experiment illustrating the same thing can be done in your bath. By moving your hand (or, if you feel energetic, your whole body) back and forth in the water at just the right frequency you can quickly set up an oscillatory motion big enough to slop water on to the floor.

Another example is that of applying a small intermittent force, at accurately judged times, to a child on a swing, by which you can work him up to quite a big amplitude of motion.

In each of these examples we find that a vibrating system responds very strongly to a force impressed from outside if this impressed force varies with time in just the right way. This strong response is called resonance, and we shall study the mathematics of resonance in this unit.

Definition 1
* *

The mathematical investigation of resonance will make use of several ideas that we have met earlier in the course; but before we can use any mathematics at all it is necessary to set up a mathematical representation, or model, of the physical situation we are studying. This mathematical model takes the form of a differential equation, but since this equation is of the second order, the techniques discussed in *Unit 24, Differential Equations I* cannot be applied directly to yield an exact (or formula) solution. In fact, since there is no deductive method that gives exact solutions for all differential equations, we are really in a "problem-solving" situation of the type discussed by G. Polya, *How to Solve It**: it is necessary to marshal whatever information appears to be relevant and use it to build up a solution to the new type of differential equation. Finally we shall have to examine the solutions we have found and interpret them to obtain an explanation of the physical phenomenon of resonance with which we started.

* G. Polya, *How to Solve It*, Open University ed. (Doubleday Anchor Books 1970). This book is the set book for the Mathematics Foundation Course; it is referred to in the text as *Polya*.

31.1.1 Setting Up a Model of Resonance

31.1.1

Main Text
* *

The first step in setting up a mathematical model for resonance is to simplify the physical situation; then we can more easily see how to describe this situation in mathematical language. The essentials of the physical situation are:

(i) a system that can oscillate even if it is left undisturbed following an initial disturbance (in the experiments suggested earlier, this is the object on a string, or the water in the bath),

and

(ii) an external to-and-fro disturbance to this system (in the experiments suggested, this disturbance is the motion of your hand).

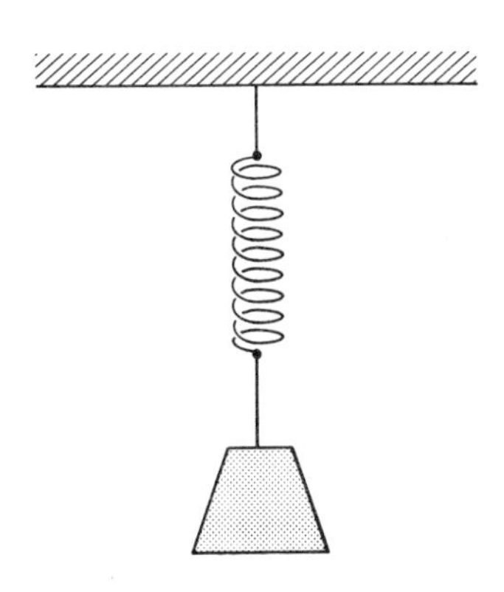

One of the simplest physical systems of this type is the one illustrated in the figure. It consists of a body hanging from a support. If the support is held fixed, the body can oscillate up and down, thus satisfying condition (i) above, and by pushing the body alternately up and down (either by hand or by moving the support up and down) we can apply an oscillatory disturbing force to the system, in accordance with condition (ii) above. The system pictured in the figure is not as artificial as it looks: the support could represent one of the axles of a car, the spring one of the car's springs, and the body the part of the car which is supported by that spring. (This is an over-simplification, of course, because we cannot divide the whole car into four independently moving parts, but it serves to illustrate the way one might set about a preliminary study of the vibrations of a car.)

In the television programme we consider a similar system in which the motion is horizontal instead of vertical. The analysis of the two systems is very similar and leads to the same differential equation.

31.1.2 Description of the Motion

31.1.2

Discussion

In setting up a mathematical representation or model of this physical system we are particularly interested in the motion of the body in the vertical direction. The mathematical model consists of two parts: (i) a description of the motion, and (ii) some mathematical equation or equations representing the physical law that determines the motion. So first we shall consider the description of the motion — that is, of how the position of the body depends on time. This is mainly a question of choosing a suitable notation. You are strongly recommended to read the articles "Notation" and "Setting up equations" on pages 134–141 and 174–177 respectively of *Polya.*

To simplify the situation as much as possible, suppose the body is concentrated in a very small region of space, so that it can be represented approximately as a mathematical point. An object so represented is called a particle. Since we are studying vertical motion, we assume the particle to move always along a vertical line. We can describe the position of the particle by giving its directed distance from some fixed point (the origin of co-ordinates) on the line. This directed distance is called the displacement of the particle; we shall denote it by the variable x.

Definition 1

Definition 2

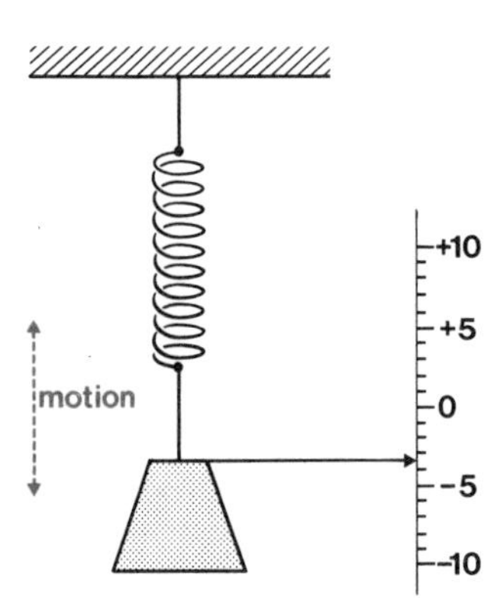

By *directed distance* we mean that the distances on one side of the origin are taken as positive, and those on the other side are taken as negative. It does not matter which direction we choose to call positive, but it is essential to make a choice and to stick to it consistently. We shall make the (completely arbitrary) choice that the upward direction is positive (i.e. points above the origin have positive values of x). Likewise, it does not matter which point we choose as origin. In oscillation problems, however, it is usually convenient to choose the origin to be the *equilibrium position* of the particle — that is, the position at which the particle can remain at rest if undisturbed.

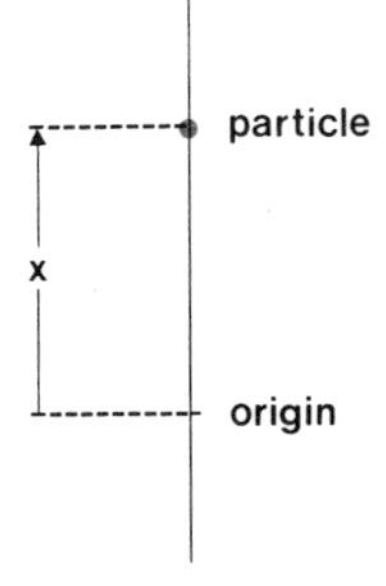

In principle, the motion can be described by giving the position of the particle at every instant of time. Thus the motion is, in principle, described by a list of ordered pairs of the form

(*time*, *position*);

in other words, by a real function

$$X: time \longmapsto displacement.$$

The graph of this function is called the *displacement-time graph*. The following diagram shows a possible displacement-time graph for an oscillating system.

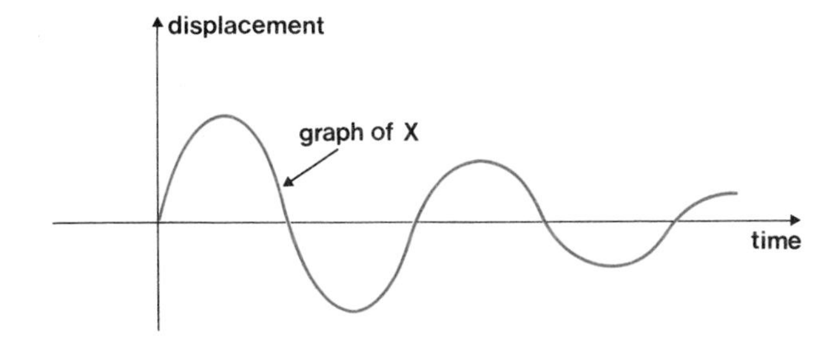

To measure the time, we once again choose an origin, that is, a particular instant from which time is to be measured. The instant at which the motion begins is often the most convenient origin: with this choice we can define any instant during the motion by giving the time (in, say, seconds) that has elapsed since the motion began. This elapsed time is usually denoted by the variable t. In this way, the motion is described by a real function

$$X: t \longmapsto x \qquad (t \in R_0^+),$$

so chosen that the displacement x at time t is $X(t)$ i.e.

$$x = X(t).$$

We have followed the normal practice of taking the domain of this function to be R_0^+, although strictly it would be more correct to use the interval $[0, T]$, where T is the total duration of the conditions determining the particular type of motion we are studying. By choosing R_0^+ as domain we are making the simplifying assumption that these conditions persist for ever.

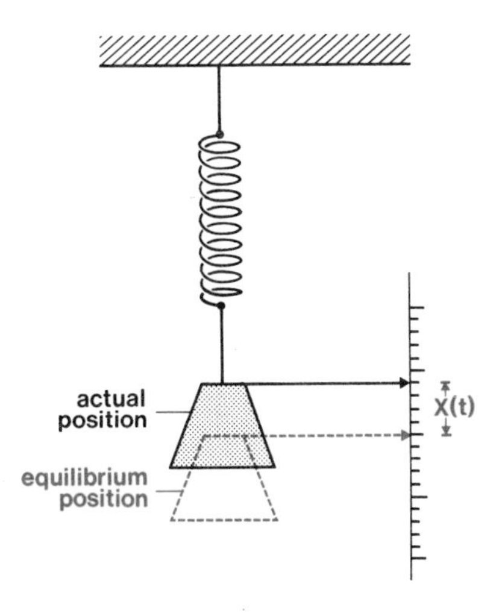

Exercise 1

Exercise 1
(2 minutes)

Express the velocity and the acceleration (rate of change of velocity) of the particle at time t in terms of the derived functions of X. ■

Exercise 2

Exercise 2
(4 minutes)

What description, analogous to the one in the text, would be suitable for the motion of a rotating piece of machinery, such as the rotor of a lathe? ■

31.1.3 Applying the Laws of Mechanics

31.1.3

Main Text
* *

Having decided to describe the motion using the function X, the next step is to find out as much as we can about this function. Mathematics alone tells us nothing about the function; we must appeal to our non-mathematical knowledge about the system under study.

Here this knowledge consists of laws of mechanics that have been discovered experimentally. One of the principal laws in the science of mechanics is Newton's second law, which states that the acceleration of any given particle is proportional to the total force acting on it. (We shall not enter here into the tricky question of how to define *force*.) In terms of our description of the motion, Newton's second law can be written

$$X''(t) = kF(t),$$

where k is the constant of proportionality and $F(t)$ is the total force acting on the particle at time t. The heavier the particle, the less it accelerates under a given force, and so the smaller k: consequently the reciprocal $\frac{1}{k}$ is larger for heavier particles. This reciprocal is denoted by m and called the mass of the particle. Its numerical value for a given particle depends on the units of measurement used. The standard international unit of mass is the kilogram. (In the television programme we choose the unit of mass to be the mass of the metal block, so that $m = 1$ there.)

Definition 1
* *

We can put the above differential equation in the form

$$mX''(t) = F(t).$$

This equation is the usual form of Newton's second law of motion, but it does not yet tell us anything about the motion; for this we need to know what is meant by the total force $F(t)$ on the particle — or at least how to calculate it. The "total force" referred to in Newton's law means the sum of the separate forces acting on the particle, and in the present case these include the force of gravity and the force exerted by the spring. For a first look at the behaviour of this system we shall assume that these are the *only* forces acting on the particle. This implies, in particular, that we ignore friction and also (for the time being) deny the possibility of "external" forces such as someone pushing the weight. The motion of a vibrating system which is not affected by any forces (apart from gravity) arising from outside itself is called free vibration to distinguish it from motion where an external force (of a particular kind) does act, which is called forced vibration.

Definition 2
* *

Definition 3
* *

Experiment shows that to a good approximation, for a small system such as that which we are considering, the force of gravity on a particle is a constant and the force of the spring varies linearly with its length (Hooke's Law).

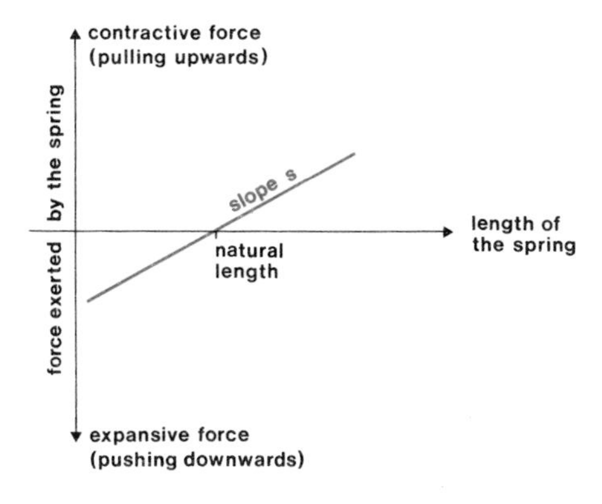

Hooke's Law

The slope of the line is called the *stiffness* of the spring, and we shall denote it here by s. In the television programme we choose the units of force (and, by implication, of time) so that $s = 1$, but here we do not make any special choice.

The total force is the sum of the spring force and the downward force due to gravity, which is independent of the length of the spring. Their sum therefore also varies linearly with the length of the spring, and the slope of the new line is again s.

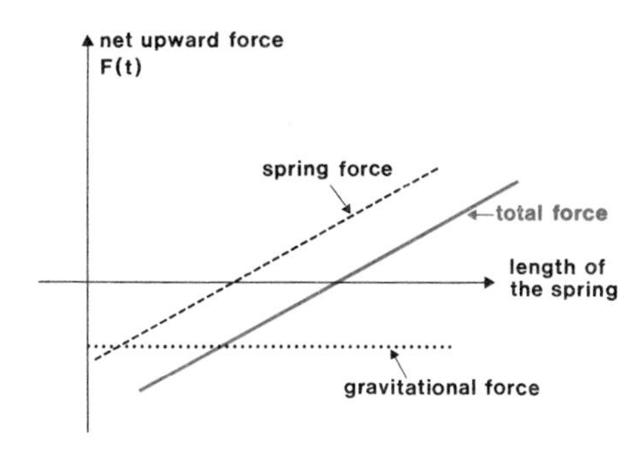

The equilibrium length of the spring is the one at which the two forces on the spring balance out; we denote this length by l_0 and the actual length by l. We can write the relation indicated in the above diagram between the net upward force $F(t)$ and the length of the spring, in the form of an equation:

$$F(t) = (l - l_0)s.$$

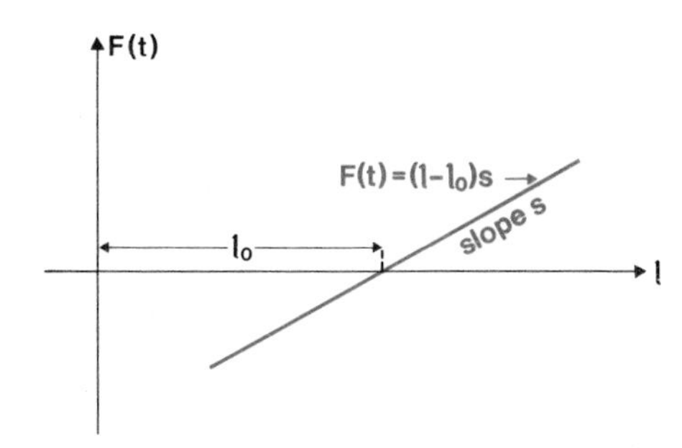

(*continued on page 9*)

Solution 31.1.2.1

Solution 31.1.2.1

The velocity is the rate of change of position, and its value at time t is therefore given by the derivative of X at t,

$$(\textit{velocity at time } t) = X'(t).$$

The acceleration is the rate of change of velocity, and its value at time t is therefore given by the second derivative of X at t,

$$(\textit{acceleration at time } t) = X''(t).$$ ■

Solution 31.1.2.2

Solution 31.1.2.2

Here, instead of a position on a line, it is the orientation of the rotor that we wish to describe. In place of the linear scale in the spring example, we have a scale measuring angles.

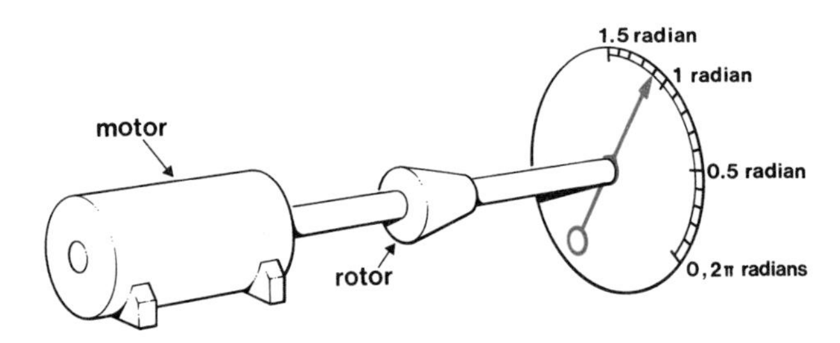

The motion can be described by a real function

$$f: t \longmapsto (\textit{reading of pointer})$$

with the pointer reading in radians. This representation, however, has the disadvantage that $f(t)$ not only jumps from 2π to 0 as the pointer passes (anti-clockwise) through zero, but also does not record the (net) magnitude of the rotation. To avoid these difficulties it is usual to work instead with the function

$$f: t \longmapsto (\textit{reading of pointer} + 2\pi n),$$

where n is the net number of times the pointer has passed zero, counting anti-clockwise as positive. That is, every time the pointer passes zero in the anti-clockwise direction, we add a further 2π to its subsequent readings, and every time it passes zero in the clockwise direction we subtract 2π. ■

(*continued from page 7*)

From the definitions of l, l_0 and $X(t)$, we have (see the following diagram)

$$l_0 = l + X(t).$$

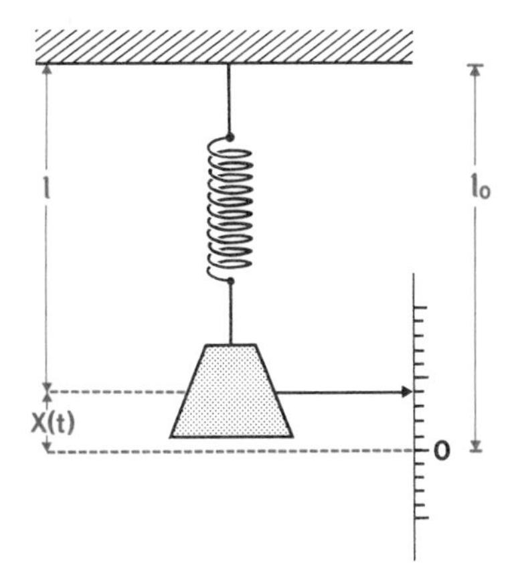

So $l - l_0 = -X(t)$ and the formula for $F(t)$ can be written

$$F(t) = -sX(t).$$

From the science of mechanics we have thus obtained two equations to help us to determine the motion of the particle in the case where the only forces are the spring force and gravity. They are

$$mX''(t) = F(t) \qquad \text{(Newton's Law)}$$

and

$$F(t) = -sX(t) \qquad \text{(Hooke's Law).}$$

These are two simultaneous equations for two unknown functions X and F. The usual method for dealing with simultaneous equations is to eliminate one of the unknowns. Here, since we are more interested in the displacement function X, which describes the motion of the particle, than in the force function F, we eliminate $F(t)$, obtaining

$$mX''(t) = -sX(t)$$

or equivalently*,

$$mX''(t) + sX(t) = 0.$$

This last equation is called the equation of motion for the system. It is a differential equation which restricts the function X describing the motion of the system, and selects from the class of all conceivable functions a much smaller class of possible functions. Remember that it is based on many physical assumptions, among which are: the support is fixed; the spring is considered to have negligible weight and to obey Hooke's law exactly; there is no friction and no external force.

Definition 4
* * *

* Many authors use the notation x for the displacement $X(t)$, $\dot{x}$ for the derivative $X'(t)$ (the velocity) and $\ddot{x}$ for the second derivative $X''(t)$ (the acceleration); in this notation the equation of motion is

$$m\ddot{x} + sx = 0.$$

31.2 SIMPLE HARMONIC MOTION

31.2

31.2.0 Introduction

31.2.0

Introduction
* *

Having set up our mathematical model for free vibrations and arrived at the equation of motion

$$mX''(t) + sX(t) = 0 \qquad (t \in R_0^+),$$

we can dispense with mechanics, and for the time being study this equation as a mathematical problem. More specifically, the problem is to find the solution set of the equation of motion—that is, the set of all possible motions of the body-and-spring system. These solutions are *functions*, not images. For this reason it is often useful to write the equation of motion as a relation connecting functions rather than images:

$$mX'' + sX = O.$$

As usual when faced with a new problem, we look at some of the problem-solving maxims given in *Polya.*

31.2.1 Finding Some Solutions

31.2.1

Technique
* *

UNDERSTANDING THE PROBLEM

What is the unknown?

A set of real functions $X : R_0^+ \longrightarrow R$.

What are the data?

$$m, s.$$

What is the condition?

The set must comprise all solutions of

$$mX'' + sX = O.$$

DEVISING A PLAN

Have you seen it before?

If you have, you will find the rest of this section very easy.

Do you know a related problem?

We have indeed seen similar problems in *Unit 24, Differential Equations I.* Our present equation is also a differential equation, but it differs from the ones studied there in that it is of the second order; that is, the highest-order derived function in the equation is the second derived function. (The equations in *Unit 24* are all of the first order.)

Here is a problem related to yours and solved before. Could you use it?

In *Unit 24* we solved a differential equation that is similar to ours, except that it is of the first order. The equation is

$$f' + 2f = O,$$

and a solution is

$$f : x \longmapsto \exp(-2x + c) \qquad (x \in R).$$

While this result does not solve the second-order equation, it may prove useful at some stage, and so it is worth keeping in mind as we search for a solution of the equation

$$mX'' + sX = O.$$

For example, it may be possible to reduce the problem of solving $mX'' + sX = O$ to the problem of solving some equation of the form $f' + \alpha f = O$, where α is a constant number.

Could you solve a part of the problem?

We could try looking for just part of the solution set instead of the whole; or, more specifically, we could look for a single function X that satisfies the condition $mX'' + sX = O$. When we have done this, we may be better placed for finding the general solution (i.e. the complete solution set).

Could you restate the problem?

There are various ways of re-writing the equation; for example,

$$X'' + \frac{s}{m}X = O$$

$$D^2X + \frac{s}{m}X = O$$

$$D^2X = -\frac{s}{m}X$$

where D^2 means $D \circ D$, i.e. $D^2 : X \longmapsto X''$.

This latter form is interesting: let us restate the problem in words. We are looking for a function X which when differentiated twice is essentially itself, except for the multiple $-\frac{s}{m}$. Or again, in other words, the operator D^2 applied to the function X produces the same effect as multiplication by the number $-\frac{s}{m}$.

Put this way, we might start to look at some of the functions whose derived functions we know, to see if any have this particular form. The process of looking would involve us in selecting a function X, differentiating it once and then differentiating again. The easiest way to do this is to extend the table of standard derived functions given in *Unit 12, Differentiation I* (page 55) as we have done below:

f	f'	f''
$x \longmapsto x^m$	$x \longmapsto mx^{m-1}$	$x \longmapsto m(m-1)x^{m-2}$
$x \longmapsto \exp x$	$x \longmapsto \exp x$	$x \longmapsto \exp x$
$x \longmapsto \ln x$	$x \longmapsto \frac{1}{x}$	$x \longmapsto -\frac{1}{x^2}$
$x \longmapsto \sin x$	$x \longmapsto \cos x$	$x \longmapsto -\sin x$
$x \longmapsto \cos x$	$x \longmapsto -\sin x$	$x \longmapsto -\cos x$

(We have omitted the domains.)

There is a further table in *Unit 12*, but this one will do for now. Three functions in this table are obvious possibilities, namely

exp, sin and cos.

None of them fits exactly, because we have not produced the factor $-\frac{s}{m}$, but we can do this by a simple modification:

f	f'	f''
$x \longmapsto \exp(ax)$	$x \longmapsto a \exp(ax)$	$x \longmapsto a^2 \exp(ax)$
$x \longmapsto \sin(ax)$	$x \longmapsto a \cos(ax)$	$x \longmapsto -a^2 \sin(ax)$
$x \longmapsto \cos(ax)$	$x \longmapsto -a \sin(ax)$	$x \longmapsto -a^2 \cos(ax)$

(See also *Unit 13, Integration II*, page 50, for a table of some standard primitive functions.)

In the case of the exponential function we would have to have $a^2 = -\frac{s}{m}$, which means that a is a complex number: this does not rule out this possibility, but involves some new concepts. So let us look at the other two possibilities. In each case

$$a^2 = \frac{s}{m} \quad \text{so that} \quad a = \pm\omega, \text{ where } \omega = \sqrt{\frac{s}{m}},$$

and we have a few possible solutions of our differential equation:

$$\left.\begin{array}{l} X : t \longmapsto \cos(at) \\ \text{or} \quad X : t \longmapsto \sin(at) \end{array}\right\} \quad (t \in R_0^+),$$

where $a = \pm\omega$ and $\omega = \sqrt{\frac{s}{m}}$.

(Since $\cos(-at) = \cos(at)$, we have three solutions, not four.)

Can you check the result?

Having arrived at these solutions by various exploratory steps, we should now check that these functions do satisfy the original equation. We leave you to do this, by direct substitution in the differential equation. Another interesting check is to note that we expected to get an oscillatory motion, and that is precisely what the cosine and sine functions represent.

Exercise 1

Exercise 1
(3 minutes)

Find two solutions of the equation

$$g'' - g = O,$$

where g has domain and codomain R. ■

Exercise 2

Exercise 2
(5 minutes)

Find two solutions of the equation

$$g'' - 2g' - 3g = O,$$

where g has domain and codomain R. ■

Exercise 3

Exercise 3
(5 minutes)

We have found a few solutions to our equation

$$X'' + \frac{s}{m}X = O$$

of the particle on the spring. Are there any more? Can you suggest any more? Have you got *all* the solutions?

(We offer no solution to this exercise: the subsequent text is concerned with these questions, but you may like to think about some of these points before reading on.) ■

31.2.2 Finding the General Solution

31.2.2

Main Text
**

The analysis in the preceding section has brought us some way towards the general solution (i.e. the complete solution set) for the equation $mX'' + sX = O$ which we have been looking for, but there is still some ground to cover. We have found only a few solutions of the equation of motion, whereas for a full understanding of the free vibrations of the physical system under study we need to be sure that we know *all* the solutions.

There are various ways of going about this problem of generalizing the few solutions we know to obtain the general solution (i.e. the complete solution set). For example, we could try to use the ideas described in "Generalization", *Polya*, page 108. This would set us on a road that would lead to the answer to our problem. (You may like to try the problem using these ideas before proceeding.) There is, however, a more powerful method available; this, too, can be related to one of Polya's articles: "Auxiliary Problem", *Polya*, page 50. As Polya states, there is no infallible method of discovering suitable auxiliary problems, although he does give two specific suggestions. There is, however, one auxiliary problem that we have already found useful in various places in the course from *Unit 3, Operations and Morphisms* onwards: find the morphism. You will remember that a morphism consists of a function h from one set to another (or to itself) together with binary operations $\circ$ and $\square$, defined on the domain and image set respectively, satisfying the condition

$$h(x \circ y) = h(x) \square h(y)$$

for all elements x and y in the domain of h. We shall apply the morphism idea in a vector space context to help us with the solution of our present problem.

This is probably not a suggestion which you would consider to be natural, because of the novelty of the ideas involved; yet this is precisely the sort of suggestion which is basic to mathematical thought. We shall try to explain the thought process involved. We are trying to solve an equation; the first thing that is of interest whenever we solve an equation is to be explicitly aware of the set on which the equation is defined. In general, equations can be cast in the form

$$f(x) = b,$$

where f is some function (occasionally it's a mapping which is not a function, but let's keep it simple), b is a known element of a codomain of this function and x (or a set of x's) is to be determined in the domain of f, which is the set on which the equation is defined. If b is in the image set of f, then a solution (or solutions) exist: otherwise they do not exist. Although each type of equation will usually require some special techniques, there are some ideas which apply to many types of equation. We wish to discover and isolate these ideas. To do this we try to determine any special properties of f and its domain.

Are there any operations (or relations) on the domain which give us a known structure? (By *structure* we mean something like a *group* or a *vector space*.) Is f a morphism for any of the obvious binary operations on the domain? We have already seen in *Units 23* and *26*, *Linear Algebra II* and *III*, how a vector space structure helps to solve equations. A group structure can also help.

To put it briefly, the differential equation we are trying to solve is "just another equation". Its solution will involve some peculiarities because it is a *differential* equation, but it will also involve some general algebraic

(*continued on page 14*)

Solution 31.2.1.1

Solution 31.2.1.1

The equation can be written as

$$D^2 g = g,$$

so that D^2 has the same effect on g as multiplication by 1. This suggests the exponential function. If we try $t \longmapsto \exp(at)$, then we find that we require $a^2 = 1$, i.e. $a = \pm 1$. Hence two solutions are

$$t \longmapsto e^t \quad (t \in R) \quad \text{and} \quad t \longmapsto e^{-t} \quad (t \in R).$$

You should verify that both these functions are solutions of the differential equation $g'' - g = O$. ■

Solution 31.2.1.2

Solution 31.2.1.2

Since the exponential function $t \longmapsto \exp(at)$ and its derived functions are all the same except for the factors a introduced at each stage, there is still hope that we can adjust the a to find some solutions. So let us try

$$g : t \longmapsto \exp(at) \qquad (t \in R);$$

then

$$\begin{aligned} g''(t) - 2g'(t) - 3g(t) &= a^2 \exp(at) - 2a \exp(at) - 3 \exp(at) \\ &= (a^2 - 2a - 3) \exp(at). \end{aligned}$$

This will be zero for all t if

$$a^2 - 2a - 3 = 0$$

i.e. if

$$(a - 3)(a + 1) = 0$$

i.e. if

$$a = 3 \text{ or } a = -1.$$

Thus two solutions are

$$g : t \longmapsto \exp(3t) \qquad (t \in R);$$

$$g : t \longmapsto \exp(-t) \qquad (t \in R).$$

You should verify that both these functions are solutions of the differential equation. ■

(continued from page 13)

principles. We have seen some of the peculiarities in the solution functions already found; we are now going to look at some of the general principles.

The ideas involved cut across subject boundaries in mathematics and are so important that we devote a television programme to their discussion. We no longer isolate an equation in the context in which it arises, but look at it from a more general point of view.

To return to the present case: we are particularly interested in the set of all functions satisfying the equation $mX'' + sX = O$. The characteristic part of our differential equation is the expression $mX'' + sX$, and so we are led to study the operator defined by this expression, that is,

$$L : f \longmapsto mf'' + sf \qquad (f \in F).$$

(The domain, F, of L will be some set of functions. We could stop to specify it more precisely, but, since we are not at present concerned with rigour, it would be a digression to do so.) We can now write our differential equation in the very concise form

$$L(X) = O,$$

but this simplified notation only helps if the relevant properties of the operator L can also be expressed in a simple form. This is where the morphism comes in.

To satisfy the definition of a morphism, we need binary operations in the domain and image set of the operator L. The most important binary operations that can be applied to functions were discussed in *Unit 1, Functions*; they are addition, multiplication and composition. Think for a moment which of these is most likely to produce a morphism when used with L (i.e. to be compatible with L). The rules of differentiation, given in *Unit 12, Differentiation I*, provide a clue: it was only for addition that we found a morphism, the one described by the rule

$$(f + g)' = f' + g'.$$

This suggests trying addition in both the domain and the image set. We find that

$$\begin{aligned} L(f + g) &= m(f + g)'' + s(f + g) \\ &= (mf'' + sf) + (mg'' + sg) \end{aligned}$$

or, in other words,

$$L(f + g) = L(f) + L(g) \qquad (f, g \in F),$$

so that there is indeed a morphism

$$L:(F, +) \longmapsto (L(F), +)$$

associated with the operator L.

It would be possible to apply the morphism at once to the solution of our differential equation, but the work will be easier if we follow the algebraic line of thought a little further first. Although we have not specified the domain F in detail, we can without harm assume that F is a vector space for the operations of addition of functions and multiplication of functions by real numbers. The morphism property of L demonstrated above is one of the two morphism properties which characterize a morphism between vector spaces, i.e. a *linear transformation* (see *Unit 23, Linear Algebra II*); it is therefore natural to inquire whether L also satisfies the other morphism property:

$$L(\alpha f) = \alpha L(f) \qquad (\alpha \in R \text{ and } f \in F).$$

Using the rules of differentiation, we can verify that this equation does indeed hold, since

$$L(\alpha f) = m(\alpha f)'' + s(\alpha f) = \alpha(mf'' + sf) = \alpha L(f).$$

Thus the operator L has two properties:

$$L(f + g) = L(f) + L(g)$$
$$L(\alpha f) = \alpha L(f)$$

and therefore satisfies the definition of a vector space morphism; that is, L is a linear transformation.

Having shown that our differential equation can be written in the form

$$L(X) = O,$$

where L is a linear transformation, we can use the theory of linear transformations discussed in *Unit 23, Linear Algebra II* to help solve this equation. In fact, we have already studied equations of this form in *Unit 23*. We called the solution set the *kernel* of the linear transformation, and we saw that it had some remarkable properties:

(i) the kernel is itself a vector space (a subspace of F);
(ii) dimension of $L(F)$ = (dimension of F) − (dimension of kernel).

The second property does not help us directly (we have not specified F), but the first is very valuable indeed: it tells us that the solution set of our differential equation is a vector space. Expressed in plain language, this means that if X and Y are any two solutions of the differential equation, then any linear combination of X and Y is also a solution of the differential equation.

We are now very close to the solution of our problem, finding the solution set of the differential equation $mX'' + sX = O$. We know a few "solutions" of this equation, namely the functions

$$\left.\begin{aligned} X &: t \longmapsto \cos(at) \\ X &: t \longmapsto \sin(at) \end{aligned}\right\} \quad (t \in R_0^+),$$

where

$$a = \pm\omega \quad \text{and} \quad \omega = \sqrt{\frac{s}{m}}$$

and we also know that the solution set is a vector space. How can these known solutions lead us to a specification of the entire solution set? We saw in *Unit 23, Linear Algebra II* that the most convenient way to specify a vector space is by giving a *basis* of the space, that is, a set of *base vectors* in it such that each element of the vector space can be expressed uniquely as a linear combination of the base vectors. It is natural to guess that perhaps the solutions we already know form a basis for the vector space constituting our solution set — in other words, that every solution of the differential equation is a unique linear combination of the solutions we have already found. There is one snag: a basis is made up of *linearly independent* elements, and of the solutions we have found only two are linearly independent, because

$$\cos(-\omega t) = \cos(\omega t)$$

and

$$\sin(-\omega t) = -\sin(\omega t).$$

So we modify our guess a little and suggest that the solution set of $mX'' + sX = O$ is the set of all functions of the form

$$X : t \longmapsto \alpha \cos \omega t + \beta \sin \omega t \qquad (t \in R_0^+),$$

where α and β are arbitrary real numbers, and $\omega = \sqrt{\frac{s}{m}}$.

Exercise 1

Exercise 1
(3 minutes)

Verify by substitution in the differential equation $mX'' + sX = O$ that every function of the above form is a solution of this differential equation.

■

Exercise 2

Exercise 2
(4 minutes)

We have taken it for granted that the two functions we are using for our basis are linearly independent. Can you prove that they are? (Remember that two vectors $\underline{v}_1$ and $\underline{v}_2$ are linearly independent if the condition

$$\alpha\underline{v}_1 + \beta\underline{v}_2 = \underline{0}$$

implies that the real numbers α and β are both 0.) ■

31.2.3 A Theorem

31.2.3

Main Text
* * *

There are still some loose ends to be tied up before we can convincingly claim to have found the general solution of our equation $mX'' + sX = O$. What we have done is to find, by a process of enlightened guesswork, or, as Polya calls it, "heuristic reasoning", a set of solutions of the equation. (See "Heuristic reasoning", *Polya*, page 113.) Since there is no obvious way of enlarging the set further, it is natural to guess that this set is in fact the set of *all* solutions. Heuristic reasoning is good in itself, but it is bad to confuse heuristic reasoning with rigorous proof. Another way of putting it is that we have solved the "problem to find" but we are still left with a "problem to prove". (See "Problems to find, problems to prove", *Polya*, pages 154–156.)

Since the Foundation Course is more concerned with showing you the main concepts in mathematics than with rigorous proofs, we shall not go into the technique for finding such proofs here; instead we quote a theorem from which the proof we require can be quickly obtained. This theorem applies to a class of operators which includes the specific operator $L: f \longmapsto mf'' + sf$ which we have been considering. The operators in this class are called linear differential operators; they are operators of the form

Definition 1
* * *

$$L: f \longmapsto k_n \times D^n f + k_{n-1} \times D^{n-1} f + \cdots + k_1 \times Df + k_0 \times f \qquad (f \in F_n),$$

where n is a positive integer called the order of the operator, F_n is a suitable set of functions, and $k_0, k_1, \ldots, k_n$ are real functions with the same domain as the functions in F_n. (More precisely, F_n is the set of all functions f, with codomain R and domain some subset of R, for which the nth derived function is continuous and has the same domain as f; the functions $k_0, \ldots, k_n$ are also required to be continuous.) We specify that k_n is not the zero function (i.e. that its image set contains numbers other than zero); otherwise the term $k_n \times D^n f$ would be zero, indicating that we had used too large a value of n in writing down the formula. This requirement is introduced to prevent such perversities as writing the operator $f \longmapsto f'$, whose order is 1, as $f \longmapsto 0 \times f'' + f'$, which would apparently be of order 2.

Definition 2
* *

As an example, the operator $f \longmapsto mf'' + sf$ is a linear differential operator, with $n = 2$,

$$\left.\begin{aligned} k_2 &: t \longmapsto m \\ k_1 &: t \longmapsto 0 \\ k_0 &: t \longmapsto s \end{aligned}\right\} \qquad (t \in R_0^+)$$

and F_2 the set of real functions with domain R_0^+ and continuous second derived functions with the same domain. In this case the functions $k_n, \ldots, k_0$ are all constant functions. Whenever all the k's are constant functions, L is called a linear differential operator with constant coefficients, and the functions $k_n, \ldots, k_0$ can be treated as numerical multipliers rather than functions. We note (without proof) the following properties:

Definition 3
* *

(1) F_n is a vector space for the usual operations;

(2) any linear differential operator L of order n is a vector space morphism (linear transformation) of F_n to $L(F_n)$.

(There is nothing essentially difficult about the proofs of these statements, but they are tedious and not very instructive.)

(*continued on page 18*)

Solution 31.2.2.1

Solution 31.2.2.1

If

$$X(t) = \alpha \cos \omega t + \beta \sin \omega t,$$

then differentiation gives

$$X'(t) = -\alpha\omega \sin \omega t + \beta\omega \cos \omega t$$

and

$$X''(t) = -\alpha\omega^2 \cos \omega t - \beta\omega^2 \sin \omega t$$
$$= -\omega^2 X(t),$$

so that

$$mX'' + sX = O.$$ ■

Solution 31.2.2.2

Solution 31.2.2.2

To prove that the cosine and sine functions we are using are linearly independent, we have to show that

$$\alpha \cos + \beta \sin = O \Rightarrow \alpha = \beta = 0,$$

i.e.

$$\alpha \cos \omega t + \beta \sin \omega t = 0 \qquad \text{for all } t \in R_0^+$$
$$\Rightarrow \alpha = \beta = 0.$$

One way to prove this is to consider a few special values of t. If we take $t = 0$, the equation to be satisfied reduces to $\alpha \times 1 + \beta \times 0 = 0$; i.e. $\alpha = 0$: if we take $t = \pi/2\omega$, it reduces to $\alpha \times 0 + \beta \times 1 = 0$; i.e. $\beta = 0$. ■

(*continued from page 17*)

Exercise 1

Exercise 1
(3 minutes)

Which of the following operators satisfy the definition of a linear differential operator?

(i) $H: f \longmapsto D^4 f - f^2$
(ii) $H: f \longmapsto D^4 f - f$
(iii) $H: f \longmapsto D^4 f - (x \longmapsto 2)$

In each case the domain is F_4. ■

Exercise 2

Exercise 2
(5 minutes)

Choose one of the operators from Exercise 1 which is not a linear differential operator, and show by means of a counter-example that it is also not a linear transformation. ■

We are now in a position to state an important theorem.

THEOREM

Theorem
* * *

The dimension of the kernel of a linear differential operator with constant coefficients is equal to the order of the operator.

Another way of stating the theorem is this: the solution set of an nth order differential equation of the form $Lf = O$, where L is a linear differential operator with constant coefficients, is a vector space of dimension n. The theorem also holds for a wide class of cases where L does not have constant coefficients, but we do not need this generalization here. The proof of the theorem is beyond the scope of this course, but we plan to include it in the M201 course on Linear Mathematics. If you would like to see a partial proof now, you can find one in Kreider et al., *An Introduction to Linear Analysis*, 106–108 (see Bibliography).

Main Text
* * *

Using this theorem, we can now complete our solution of the differential equation $mX'' + sX = O$.

The solution set of this equation is just the kernel of the linear differential operator $f \longmapsto mf'' + sf$, which is of second order. By the theorem, therefore, the dimension of this kernel is 2: the solution set of $mX'' + sX = O$ is a vector space of dimension 2. To specify this vector space, all we need is a basis, that is, a set of 2 linearly independent solutions.

We know already that the 2 functions

$$\left.\begin{array}{l} t \longmapsto \cos \omega t \\ \text{and} \\ t \longmapsto \sin \omega t \end{array}\right\} \qquad (t \in R_0^+)$$

are solutions, and that they are linearly independent (Exercise 31.2.2.2). Consequently they form a basis for the solution space. The solution space itself is the set of all linear combinations of the base vectors, i.e. the set of all functions X such that

$$X : t \longmapsto \alpha \cos \omega t + \beta \sin \omega t \qquad (t \in R_0^+),$$

where α and β are arbitrary real numbers, and $\omega = \sqrt{\frac{s}{m}}$.

Exercise 3

Exercise 3
(4 minutes)

Use the method we have used for the equation $mX'' + sX = O$ to find all solutions of the differential equation

$$D^2 f - f = O,$$

where f has domain R.

That is,

(i) determine the dimensionality of the solution set (a vector space), using the theorem;
(ii) find enough linearly independent solutions of the equation to form a basis for this vector space;
(iii) write down a formula giving the general solution in terms of these linearly independent solutions. ■

Exercise 4

Exercise 4
(4 minutes)

Find the general solution of

$$f'' - 5f' + 4f = O,$$

where f has domain R. (Cf. Exercise 31.2.1.2.) ■

Exercise 5

Exercise 5
(4 minutes)

What goes wrong when we apply the method of the preceding exercise to the equation

$$f'' + 2f' + f = O,$$

where f has domain R? ■

(continued on page 21)

Solution 1

Solution 1

(i) This operator is not linear: the f^2 term spoils it.
(ii) This operator is linear.
(iii) This operator is not linear: the $x \longmapsto 2$ term spoils it. ■

Solution 2

Solution 2

(i) With $f: x \longmapsto 1$, for example, we have

$$2H(f) = 2(D^4 f - f^2) = x \longmapsto -2$$

but

$$H(2f) = D^4(2f) - (2f)^2 = x \longmapsto -4.$$

(iii) With $f: x \longmapsto 1$ we have

$$2H(f) = 2(D^4 f - (x \longmapsto 2)) = x \longmapsto -4$$

but

$$H(2f) = D^4(2f) - (x \longmapsto 2) = x \longmapsto -2. \quad ■$$

Solution 3

Solution 3

(i) The equation is of the form $Lf = O$, where L is a linear differential operator of degree 2 with constant coefficients. By the theorem, its kernel is therefore a vector space of dimension 2, and this kernel is the solution set of the equation.
(ii) It was shown in Solution 31.2.1.1 that $D^2 f = f$ has the solutions

$$t \longmapsto e^t \qquad (t \in R)$$

and

$$t \longmapsto e^{-t} \qquad (t \in R).$$

These solutions are linearly independent. (If $\alpha e^t + \beta e^{-t} = 0$ for all t, then we have, putting $t = 0$ and then $t = 1$, $\alpha + \beta = 0$ and $\alpha e + \beta e^{-1} = 0$, and these two equations taken together imply that $\alpha = \beta = 0$.)
(iii) Taking the two linearly independent solutions given in (ii) as base vectors for the vector space, we obtain the general solution:

$$\alpha(t \longmapsto e^t) + \beta(t \longmapsto e^{-t}) \qquad (t \in R),$$

i.e.

$$t \longmapsto \alpha e^t + \beta e^{-t} \qquad (t \in R),$$

where α and β are arbitrary real numbers. ■

Solution 4

Solution 4

If we try

$$f: t \longmapsto \exp(ct) \qquad (t \in R)$$

in the equation, the left-hand side becomes

$$f'' - 5f' + 4f = t \longmapsto (c^2 e^{ct} - 5ce^{ct} + 4e^{ct}) \qquad (t \in R)$$
$$= (c^2 - 5c + 4)(t \longmapsto e^{ct}) \qquad (t \in R).$$

This equation is satisfied if the right-hand side reduces to the zero function, i.e. if

$$c^2 - 5c + 4 = 0$$

i.e.

$$c = 1 \text{ or } 4.$$

Thus, the differential equation has the two solutions

$$t \longmapsto \exp t \quad \text{and} \quad t \longmapsto \exp 4t,$$

which can be verified to be linearly independent. Using them as base vectors for the two-dimensional vector space of solutions, we obtain the general solution:

$$t \longmapsto \alpha \exp t + \beta \exp 4t \qquad (t \in R),$$

where α and β are arbitrary real numbers. ■

Solution 5

Solution 5

Applying the same method as before, we find that the condition for $t \longmapsto \exp(ct)$ to be a solution is

$$c^2 + 2c + 1 = 0.$$

This quadratic equation has only one solution, $c = -1$, and so the differential equation has only one solution of the form $t \longmapsto \exp(ct)$, namely

$$t \longmapsto \exp(-t) \qquad (t \in R).$$

We know, however, that the solution space is two-dimensional; therefore the single function displayed above is insufficient for a basis for the solution space. In this case, therefore, the method we have been using is inadequate. The difficulty can be overcome: you may like to have a try. ■

(continued from page 19)

To end this section we give an example of how some formalistic manipulation can be put to good use.

Example 1

Example 1

If we try to solve the equation

$$f'' + 4f' + 5f = O$$

by the method described in the text, we find that $t \longmapsto \exp(ct)$ is a solution provided that

$$c^2 + 4c + 5 = 0$$

i.e.

$$c = -2 \pm i.$$

This could be considered to be a little embarrassing for at least two reasons:

(i) we have always assumed that our functions were real functions;

(ii) we have never discussed differentiating functions which are not real functions, so even if we assume that we have found two solutions, we have no means of verification.

We might, therefore, conclude that if the equation has (real) solutions, they are not of exponential form. That is strictly correct, but there is no need to throw out the baby with the bath water. In *Unit 29, Complex Numbers II* we came across Euler's formula:

$$e^{i\phi} = \cos \phi + i \sin \phi.$$

Using this, we can write

$$e^{(-2+i)t} = e^{-2t}e^{it} = e^{-2t}(\cos t + i \sin t)$$

and

$$e^{(-2-i)t} = e^{-2t}e^{-it} = e^{-2t}(\cos t - i \sin t).$$

Now even if the differentiation of the complex exponential function is in doubt, the algebra is still valid: the vector space spanned by the complex exponential functions $t \longmapsto \exp(-2+i)t$, $t \longmapsto \exp(-2-i)t$, can be spanned by any other two linearly independent vectors in the same space. And we notice that

$$e^{(-2+i)t} + e^{(-2-i)t} = 2e^{-2t} \cos t$$

$$e^{(-2+i)t} - e^{(-2-i)t} = 2ie^{-2t} \sin t.$$

Forgetting about the 2 and $2i$ (which are just scalar multipliers, the set of scalars in this case being the set of complex numbers), we see that

$$t \longmapsto e^{-2t} \cos t \quad \text{and} \quad t \longmapsto e^{-2t} \sin t$$

belong to the same two-dimensional vector space. They can be shown to be linearly independent and hence form a basis for that vector space. So, using "heuristic reasoning", we may conjecture that the solution set of the equation is

$$t \longmapsto \alpha e^{-2t} \cos t + \beta e^{-2t} \sin t \qquad (t \in R),$$

where α and β are arbitrary real numbers. We leave you to verify by substitution that this is correct. The linear independence then means that we have found the set of all real solutions of our equation. ■

Discussion

* *

This example illustrates what turns out to be a valid method of solving such equations. It is interesting to note that these differential equations are not the figments of the imagination of a demented mathematician. Such equations occur in electric circuit theory, for instance. The e^{-2t} has a "damping" effect on the rest of the solution: as t increases so e^{-2t} decreases, and for "large" t the solution becomes effectively

$$t \longmapsto 0 \qquad (t \in R).$$

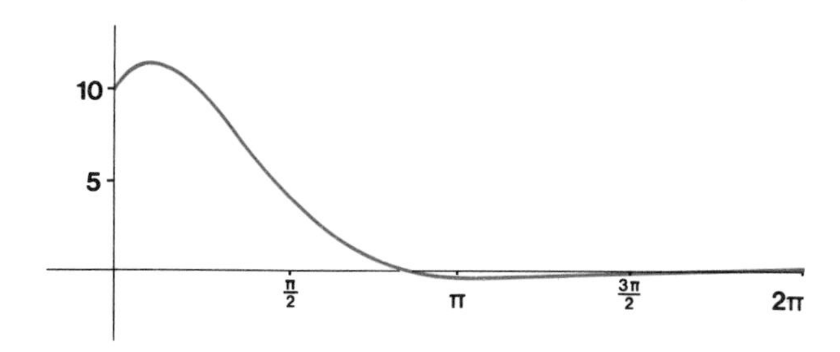

The above function would arise if, for instance, a system were displaced from its equilibrium position and then left to its own devices without further outside interference. The initial displacement would set up oscillations which would gradually die away, in the same way as the oscillations set up by dropping a stone in water.

31.2.4

Discussion
★ ★

31.2.4 Interpretation of the Solution

In this section we interpret the general solution,

$$X: t \longmapsto \alpha \cos \omega t + \beta \sin \omega t \qquad (t \in R),$$

which we have found for the equation of motion of our body-spring system, in terms of the possible motions of the system. Every solution of the equation with amplitude less than l_0 corresponds to a possible motion, and so the general solution we have found, with this restriction on α and β, corresponds to the most general possible motion. We can use it to find the general properties of the motion of the vibrating system.

The most important of these properties is that the motion represented by our solution is *periodic*: it repeats itself exactly again and again. For example, here are the graphs of the two solutions we have been using as base vectors of our vector space of solutions.

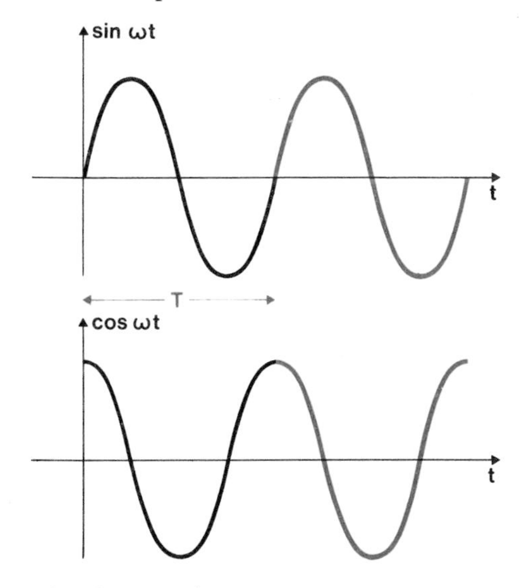

Here T is defined by the equation

$$\omega T = 2\pi$$

i.e.

$$T = \frac{2\pi}{\omega},$$

so that the first function, $t \longmapsto \sin \omega t$, maps T to $\sin 2\pi$ and the second, $t \longmapsto \cos \omega t$, maps T to $\cos 2\pi$. In both the graphs the section from 0 to T is exactly repeated in the sections from T to $2T$, $2T$ to $3T$, and so on. Just the same thing happens for linear combinations of the two functions; for example:

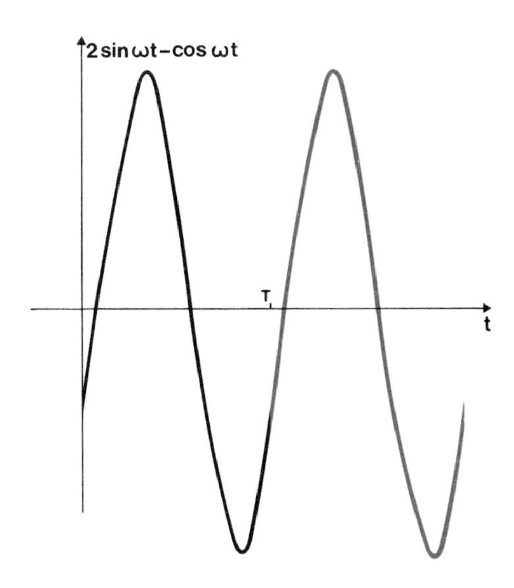

The number T is called the period of the motion. In this case T depends only on the mass of the body and the stiffness of the spring, and not on the particular motion of the vibrating system. The fact that the period remains constant is of the utmost importance for vibrating systems. It ensures, for example, that the pitch of the note emitted by a tuning fork, which is determined by the period of the oscillations of the tines, is not affected by the amplitude (the "size") of the vibrations — a very desirable prerequisite for tuning forks to fulfil their function of producing notes which serve as a standard of pitch.

Definition 1
**

Another feature of the general solution which we can relate to the properties of the mechanical system is the presence of two unspecified numbers α and β, usually called the arbitrary constants, in the solution

Definition 2
**

$$X: t \longmapsto \alpha \cos \omega t + \beta \sin \omega t \qquad (t \in R).$$

For *first*-order differential equations, we found in *Unit 24, Differential Equations I* that the general solution contained only *one* arbitrary constant. For *second*-order differential equations of the linear type (which is the only type considered here), the general solution has *two* arbitrary constants, and in fact an nth-order linear differential equation normally has n arbitrary constants in its general solution.

To pick out, from the family of solutions (the vector space) the particular solution describing some particular motion of the mechanical system, we need two pieces of information to determine the two numbers α and β. In mechanical problems such as the body-spring system this information often concerns the initial position and velocity of the body; remember that $X(t)$ is its displacement from the equilibrium position at time t, and $X'(t)$ is its velocity at time t. For example, if the system is started at time 0 from rest at the position x_0, then we have

$$X(0) = x_0, \text{ since the initial position is } x_0,$$

and

$$X'(0) = 0, \text{ since the initial velocity is } 0.$$

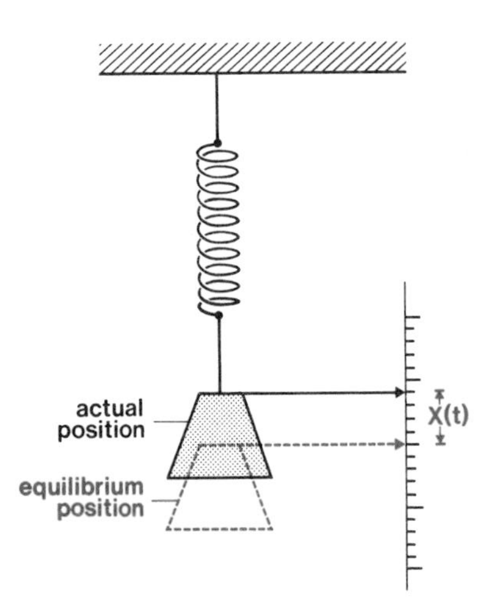

These two pieces of information can be used to fix the values of α and β; since in this case, $X(0) = \alpha$ and $X'(0) = \beta$, we obtain $\alpha = x_0$, $\beta = 0$. So the solution, if the system starts from rest at a position x_0, is given by

$$X: t \longmapsto x_0 \cos \omega t \qquad (t \in R_0^+).$$

Exercise 1

Exercise 1
(3 minutes)

If the particle starts at time 0 from the position x_0 with velocity v_0, find the appropriate solution which describes the displacement. ■

Exercise 2

Exercise 2
(5 minutes)

If you wanted to increase the frequency (the number of oscillations per second) of a vibrating system would you
(a) increase the mass;
(b) stiffen the spring;
(c) decrease the mass;
(d) make the spring less stiff? ■

31.3 FORCED VIBRATIONS AND RESONANCE

31.3

31.3.1 A Mathematical Model for Resonance

31.3.1

Discussion
* *

To deal mathematically with the phenomenon of resonance, we must extend our mathematical model by including the effects of forces other than the force $-sX(t)$ that arises directly from the displacement of the body. Vibrations in the presence of such additional forces are called forced vibrations, to distinguish them from free vibrations which take place in the absence of such forces. In studying resonance we are particularly interested in the case where this additional force is periodic: we shall see how the oscillations excited by such a force can build up to a very large amplitude even when the amplitude of the force itself is small. One way in which such an additional force could be transmitted to the system would be for the support of the spring to move in periodic fashion.

Definitions 1
* *

For our mathematical treatment of this physical situation, let us take the additional force to have the form $F_0 \cos pt$, where F_0 and p are numbers which we can use to adjust its amplitude and period. (In the television programme we use a sine instead of a cosine.) The reason for using a cosine or a sine instead of some other periodic function is that the equation of motion is easily solved for a cosine or sine. When this new force is taken into account, Newton's second law gives

$$\begin{aligned} mX''(t) &= \textit{total force on body} \\ &= -sX(t) + F_0 \cos pt \qquad (t \in R_0^+). \end{aligned}$$

This is usually written with the unknown function X taken to the left-hand side:

$$mX''(t) + sX(t) = F_0 \cos pt \qquad (t \in R_0^+).$$

This equation of motion, with a particular choice of p, is the basis of our study of resonance. After solving the general case as a mathematical problem, we shall interpret its solution in terms of the behaviour of the physical system it represents.

(*continued on page 26*)

Solution 31.2.4.1

Solution 31.2.4.1

In this case

$$X(0) = x_0 \quad \text{and} \quad X'(0) = v_0,$$

so that

$$X : t \longmapsto x_0 \cos \omega t + \frac{v_0}{\omega} \sin \omega t \qquad (t \in R_0^+). \qquad \blacksquare$$

Solution 31.2.4.2

Solution 31.2.4.2

The number of oscillations per second is $1/T$, which is equal to $\omega/2\pi$. To increase this, alternatives (b) and (c) (increase s or decrease m) are appropriate.

This result can also be understood on an intuitive basis directly from the mechanics: a light mass is moved rapidly by a strong spring, whereas a heavy mass attached to a weak spring will be sluggish. ■

(*continued from page 25*)

31.3.2 Solving the Equation of Motion

31.3.2

Technique
* *

Faced with the problem of solving the equation of motion

$$mX''(t) + sX(t) = F_0 \cos pt \qquad (t \in R_0^+),$$

we can once again turn for help to Polya's list of questions.

UNDERSTANDING THE PROBLEM

What is the unknown? A set of real functions of the form $f : R_0^+ \longrightarrow R$.

What are the data? m, s, F_0 and p.

What is the condition? The set must comprise all solutions of the above equation.

DEVISING A PLAN

Do you know a related problem?

We have just dealt with the related problem of solving

$$mX'' + sX = O,$$

i.e.

$$mX''(t) + sX(t) = 0.$$

Here is a problem related to yours and solved before. Could you use it?

We know that the solution set of $mX'' + sX = O$ is the vector space spanned by the two functions

$$t \longmapsto \cos \omega t, \qquad t \longmapsto \sin \omega t.$$

The equation $mX'' + sX = O$ is a special case (with $F_0 = 0$) of the one we are now looking at, and so our problem is to generalize the solution of $mX'' + sX = O$ to the case where the right-hand side is a non-zero function.

Could you solve a part of the problem?

We could try looking for only a part of the solution set instead of the whole: that is, for a *particular* solution of our equation

$$mX''(t) + sX(t) = F_0 \cos pt.$$

As it happens, this equation does have a simple particular solution. Can you see a way of getting it?

Exercise 1

Exercise 1
(3 minutes)

Show that, if $p^2 \neq s/m$, then there is a number k such that

$$X: t \longmapsto k \cos pt \qquad (t \in R_0^+)$$

is a solution of

$$mX''(t) + sX(t) = F_0 \cos pt.$$ ■

Exercise 2

Exercise 2
(3 minutes)

Find a particular solution of

$$X''(t) + X(t) = \sin pt, \qquad (t \in R_0^+)$$

assuming $p^2 \neq 1$. ■

Technique
* *

Coming back to our problem, we now have a particular solution of the differential equation, but this is only a partial solution of our problem since there may be other solutions of the equation. To find them, we continue with Polya's questions.

Could you restate the problem?

The equation can be rewritten in various ways. For example, we can use the function notation instead of the image notation. This gives

$$mX'' + sX = t \longmapsto F_0 \cos pt$$

or, more concisely,

$$mX'' + sX = g$$

where $g: t \longmapsto F_0 \cos pt$. A still more concise statement would be

$$L(X) = g,$$

where L is the linear operator that maps X to $mX'' + sX$.

We have come across equations of this latter form before. In *Unit 26, Linear Algebra III*, section 26.2.2, we studied systems of linear equations of the form

$$A\underset{\sim}{x} = \underset{\sim}{b},$$

where $\underset{\sim}{x}$ and $\underset{\sim}{b}$ are column vectors and A is an $n \times n$ matrix. We found that, if A is non-singular, then $A\underset{\sim}{x} = \underset{\sim}{b}$ has a unique solution; but that if A is singular, then $A\underset{\sim}{x} = \underset{\sim}{b}$, if it has any solution at all, has many solutions. In the latter case, we can find many solutions as follows: if $\underset{\sim}{z}$ is any element of the kernel of the mapping represented by A (that is, a solution of $A\underset{\sim}{x} = \underset{\sim}{0}$) and $\underset{\sim}{x}_0$ is any solution of $A\underset{\sim}{x} = \underset{\sim}{b}$, then $\underset{\sim}{x}_0 + \underset{\sim}{z}$ is another solution of $A\underset{\sim}{x} = \underset{\sim}{b}$; for we have

$$A(\underset{\sim}{x}_0 + \underset{\sim}{z}) = A\underset{\sim}{x}_0 + A\underset{\sim}{z} = \underset{\sim}{b} + \underset{\sim}{0} = \underset{\sim}{b}.$$

It is not difficult to prove also that *every* solution of the equation $A\underset{\sim}{x} = \underset{\sim}{b}$ is of the form $\underset{\sim}{x}_0 = \underset{\sim}{z}$ with $\underset{\sim}{z}$ in the kernel. (We shall prove this later in this text.)

(*continued on page 28*)

Solution 1

Solution 1

Substituting the suggested form of X into the differential equation, we obtain the condition

$$m(-p^2k\cos pt) + sk\cos pt = F_0\cos pt$$

i.e.

$$(-mp^2k + sk)\cos pt = F_0\cos pt,$$

which is satisfied if

$$-mp^2k + sk = F_0.$$

Since $p^2 \neq s/m$, this equation can be solved for k, giving

$$k = \frac{F_0}{s - mp^2}.$$

The relevant particular solution of the equation is therefore

$$X: t \longmapsto \left(\frac{F_0}{s - mp^2}\right)\cos pt \qquad (t \in R_0^+).$$ ■

Solution 2

Solution 2

The given equation is like the previous one, but with a sine in place of a cosine. This suggests trying

$$X: t \longmapsto k\sin pt \qquad (t \in R_0^+)$$

as a possible solution, and substitution in the differential equation shows that this function does satisfy the equation, where

$$k = \frac{1}{1 - p^2}.$$

A particular solution of the equation is thus

$$X: t \longmapsto \left(\frac{1}{1 - p^2}\right)\sin pt \qquad (t \in R_0^+).$$ ■

(continued from page 27)

CARRYING OUT THE PLAN

We can now carry out the third of Polya's four main steps in the problem-solving process. We know that the general solution of the matrix equation

$$A\underset{\sim}{x} = \underset{\sim}{b}$$

is

$$\underset{\sim}{x} = (\text{some particular solution of } A\underset{\sim}{x} = \underset{\sim}{b}) + (\text{general element of the kernel of } A).$$

We also know that our situation here is effectively the same: the function X belongs to a set of functions which forms a vector space (corresponding to the vector space to which the $\underset{\sim}{x}$'s belong), and the operator L is a linear transformation on this vector space (corresponding to the linear transformation represented by A). It is therefore to be expected that we can find the general solution of the differential equation $L(X) = g$ in an analogous way:

$$X = (\text{some particular solution of } L(X) = g) + (\text{general element of the kernel of } L),$$

where the kernel of L is defined as the set of functions that are mapped to O by L. A general expression for all the elements of the kernel of a differential operator L is usually called a complementary function of L, since it "complements" the particular solution to give the general solution.

Definition 1
★ ★

Exercise 3

Exercise 3
(3 minutes)

For the case where

$$g : t \longmapsto F_0 \cos pt$$

and

$$L : X \longmapsto mX'' + sX,$$

we have just found a particular solution of $L(X) = g$, and we found the general solution of $L(X) = 0$ earlier in this text. Use this information to write down a formula for the general solution of the differential equation

$$mX'' + sX = t \longmapsto F_0 \cos pt,$$

where

$$p^2 \neq \frac{s}{m}.$$

■

LOOKING BACK

Technique
★ ★

Having discovered a method that appears to solve the problem, you may be tempted to look no further and to get on as quickly as possible with something else. It is usually a mistake to yield to this temptation: you may fail to make the most of the effort you have spent in finding a method. The supposed solution may be incomplete in some way, or even wrong; further, there may be some useful lesson to be learned from it. For this reason, the fourth and last of Polya's steps in problem-solving tells us to examine the solution obtained and, in particular, to check the result. That is, we look back at the original problem and make sure that we really have the solution — neither more nor less.

In the present case, our problem is to find the general solution (i.e. the solution set) of the differential equation:

$$mX''(t) + sX(t) = F_0 \cos pt \qquad (t \in R_0^+),$$

and we have arrived at the supposed solution:

$$X(t) = \left(\frac{F_0}{s - mp^2}\right) \cos pt + \alpha \cos \omega t + \beta \sin \omega t,$$

where α and β are arbitrary real numbers.

The following two exercises, taken together, ask you to check that this really is the general solution. In both exercises you should assume that $p^2 \neq s/m$.

Exercise 4

Exercise 4
(3 minutes)

Check that every function of the form given above is a solution of the differential equation, whatever values are taken for α and β. ■

Exercise 5

Exercise 5
(5 minutes)

Check that every real solution of the differential equation has the form given above.

The hint from *Polya* that is perhaps the most useful here is: "Here is a problem related to yours and solved before. Could you use it?" ■

Solution 3

Solution 3

The particular solution of $L(X) = g$ is

$$t \longmapsto \left(\frac{F_0}{s - mp^2}\right) \cos pt \qquad (t \in R_0^+).$$

(See Exercise 1.)

The complementary function, i.e. the kernel of L, is the general solution of $mX'' + sX = O$, which we found in sections 31.2.2 and 3 to be

$$t \longmapsto \alpha \cos \omega t + \beta \sin \omega t \qquad (t \in R_0^+).$$

Putting these together, we find the general solution of $L(X) = g$ to be

$$t \longmapsto \left(\frac{F_0}{s - mp^2}\right) \cos pt + \alpha \cos \omega t + \beta \sin \omega t \qquad (t \in R_0^+),$$

provided $p^2 \neq s/m$. ■

Solution 4

Solution 4

Differentiation of the supposed solution gives

$$X'(t) = -\left(\frac{pF_0}{s - mp^2}\right) \sin pt - \alpha\omega \sin \omega t + \beta\omega \cos \omega t,$$

$$X''(t) = -\left(\frac{p^2F_0}{s - mp^2}\right) \cos pt - \alpha\omega^2 \cos \omega t - \beta\omega^2 \sin \omega t,$$

whence

$$mX''(t) + sX(t) = \left(\frac{-mp^2F_0 \cos pt + sF_0 \cos pt}{s - mp^2}\right) = F_0 \cos pt.$$

as required. ■

Solution 5

Solution 5

We have already noted the analogy between the equation under consideration here, which has the form $L(X) = g$, and the matrix equation $A\underline{x} = \underline{b}$. The "problem related to yours" is that of proving that if $\underline{x}_0$ is a particular solution of $A\underline{x} = \underline{b}$, then the general solution of $A\underline{x} = \underline{b}$ is

$$\underline{x} = \underline{x}_0 + (\text{general element of the kernel of } A).$$

The proof of this was given in section 26.2.2 of *Unit 26, Linear Algebra III*, and we repeat it here. Let $\underline{x}_0$ be the particular solution of $A\underline{x} = \underline{b}$, and $\underline{x}$ be any other solution, so that

$$A\underline{x} = \underline{b}$$

and

$$A\underline{x}_0 = \underline{b}.$$

Subtracting, and using the fact that A is a linear transformation, gives

$$A(\underline{x} - \underline{x}_0) = \underline{0},$$

so that $\underline{x} - \underline{x}_0$ belongs to the kernel of A, i.e. $\underline{x}$ is of the form

$$\underline{x}_0 + \text{an element of the kernel of } A.$$

Thus any solution of $A\underline{x} = \underline{b}$ is of this form: in other words, the general solution has the form

$$\underline{x}_0 + (\text{general element of the kernel of } A).$$

To prove the corresponding result here we need little more than a change of notation. The given differential equation can be written as

$$L(X) = g$$

with

$$L: X \longmapsto mX'' + sX \quad \text{and} \quad g: t \longmapsto F_0 \cos pt.$$

We have seen already that one solution of this equation is the function

$$t \longmapsto \left(\frac{F_0}{s - mp^2}\right) \cos pt \qquad (t \in R_0^+),$$

which we denote by X_0. Then if X is any solution of the differential equation, we have

$$LX = g$$

and

$$LX_0 = g.$$

Since L is a linear transformation, subtraction gives $L(X - X_0) = O$, so that $X - X_0$ must belong to the kernel of L, that is, to the solution set of equation $L(X) = O$. We have seen in section 31.2.3 that this set comprises all functions of the form

$$t \longmapsto \alpha \cos \omega t + \beta \sin \omega t \qquad (t \in R_0^+);$$

therefore $X - X_0$ must be of this form, and so we have, by the definition of X_0,

$$X: t \longmapsto \left(\frac{F_0}{s - mp^2}\right) \cos pt + \alpha \cos \omega t + \beta \sin \omega t \qquad (t \in R_0^+),$$

for some α and β. Thus every solution X of $L(X) = g$ does have the form stated. ■

31.3.3 Interpretation of the Solution

31.3.3

Main Text

* *

It remains to interpret the general solution we have found for the equation

$$mX''(t) + sX(t) = F_0 \cos pt \qquad \left(p^2 \neq \frac{s}{m}\right).$$

Our solution has the form

$$X: t \longmapsto \left(\frac{F_0}{s - mp^2}\right) \cos pt + \alpha \cos \omega t + \beta \sin \omega t \qquad (t \in R_0^+).$$

As we have seen, the right-hand side is the sum of

(i) a particular solution:

$$t \longmapsto \left(\frac{F_0}{s - mp^2}\right) \cos pt \qquad (t \in R_0^+),$$

under which the image is proportional to the impressed force $F_0 \cos pt$ and contains no arbitrary constants (other than those which describe the impressed force itself);

and

(ii) a complementary function:

$$t \longmapsto \alpha \cos \omega t + \beta \sin \omega t \qquad (t \in R_0^+),$$

which does not depend on the nature of the impressed force, and contains two arbitrary constants just as in the case of free vibrations.

Because of the two arbitrary constants α and β, the expression for $X(t)$ gives a whole family of functions describing possible notions of the system, one for each pair of numbers (α, β). To choose the particular member of this family corresponding to a particular motion of the system, we need two pieces of information; these are usually the position and velocity of the vibrating body when the motion starts.

Example 1

Example 1

Suppose the particle starts from rest at the origin; that is,

$$X(0) = 0 \quad \text{(starts at origin)},$$

$$X'(0) = 0 \quad \text{(starts at rest)}.$$

The general solution gives

$$X(0) = \left(\frac{F_0}{s - mp^2}\right) \cos 0 + \alpha \cos 0 + \beta \sin 0 = \left(\frac{F_0}{s - mp^2}\right) + \alpha$$

$$X'(0) = \left(\frac{-pF_0}{s - mp^2}\right) \sin 0 - \alpha\omega \sin 0 + \beta\omega \cos 0 = \beta\omega,$$

and so using the conditions $X(0) = X'(0) = 0$, we obtain

$$\alpha = \frac{-F_0}{s - mp^2}, \qquad \beta = 0.$$

The particular solution of the equation appropriate to the given information is therefore

$$X : t \longmapsto \left(\frac{F_0}{s - mp^2}\right) \cos pt - \left(\frac{F_0}{s - mp^2}\right) \cos \omega t \qquad (t \in R_0^+).$$

The following graph illustrates the case for which $F_0 = s = m = 1, p = \frac{1}{2}$.

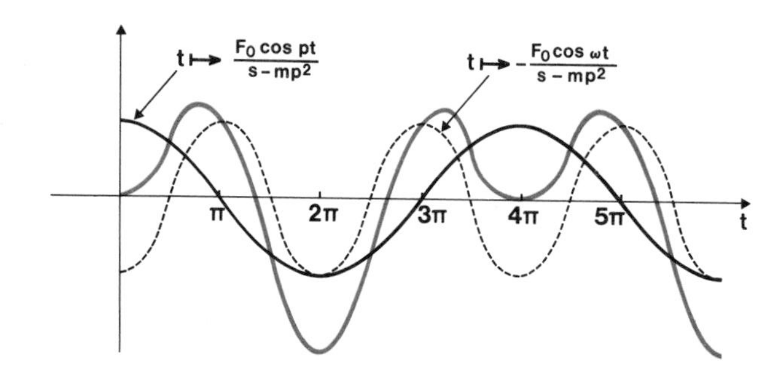

Exercise 1

Exercise 1
(4 minutes)

If the particle starts from a position x_0 with velocity v_0, find the formula for $X(t)$ which describes the subsequent motion. ■

Exercise 2

Exercise 2
(5 minutes)

If the equation of motion is the one considered in the television programme,

$$X''(t) + X(t) = \sin \tfrac{1}{2}t \qquad (t \in R_0^+),$$

and the particle starts from rest at the origin, find the formula for the displacement $X(t)$.

(See Exercise 31.3.2.2.) ■

Discussion
* *

The graphs in Example 1 and in Solution 2 show that the motion in forced vibrations is likely to be more complicated than in free vibrations. This is only to be expected, because the differential equation for forced vibrations is more complicated. The graph no longer has the character of a sinusoidal curve of period $2\pi/\omega$ as it does for free vibrations; instead, in general, it consists of a superposition of this sinusoidal wave and another, whose period is determined by the function giving the impressed force. The superposition of these two motions with different periods leads to quite complicated motions (which need not themselves be periodic).

Exercise 3
(2 minutes)

Exercise 3

(i) What is the period of the function

$$t \longmapsto \cos pt \qquad (t \in R)?$$

(ii) What is the period of the function

$$X : t \longmapsto \tfrac{4}{3} \sin \tfrac{1}{2}t - \tfrac{2}{3} \sin t \qquad (t \in R_0^+)$$

which was found in Solution 2? ■

Main Text
* *

As indicated at the beginning of this text, an important application of the mathematics of vibrations is to the phenomenon of resonance. By *resonance* we mean the tendency of the oscillations of a vibrating system to reach very large amplitudes when the force on it oscillates with a period equal, or very close, to the period of its free vibrations. Our model gives a quantitative explanation of this phenomenon. Suppose, for definiteness, that the impressed force is $F_0 \cos pt$ and that the system starts from rest at the origin at time 0. Then, by Solution 1, the subsequent motion is given by

$$X : t \longmapsto \left(\frac{F_0}{s - mp^2}\right)(\cos pt - \cos \omega t) \qquad (t \in R_0^+).$$

The periods of the free vibrations and of the applied force are $2\pi/\omega$ and $2\pi/p$. If they are close together we have p close to ω which implies that mp^2 is close to s, and so the denominator, $(s - mp^2)$, is small. Consequently, the factor $\left(\dfrac{F_0}{s - mp^2}\right)$ is large, and so $X(t)$ will be large except at those times (such as the initial time) when the two cosines cancel out, or nearly cancel out. The diagrams show graphs of X, with $F_0 = m = s = 1$, for two different values of p. The closer p is to $\omega = 1$, the larger is the maximum amplitude.

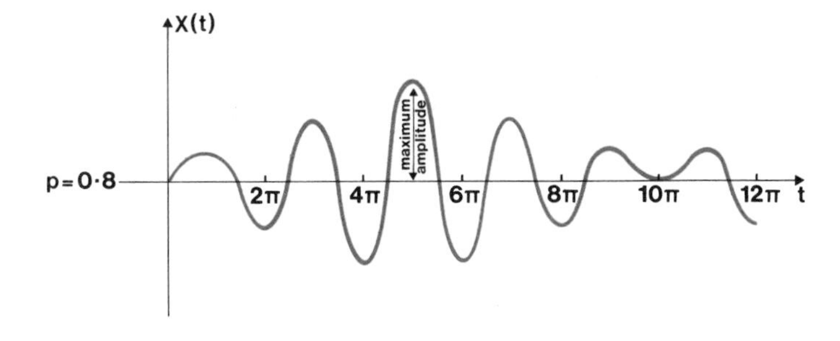

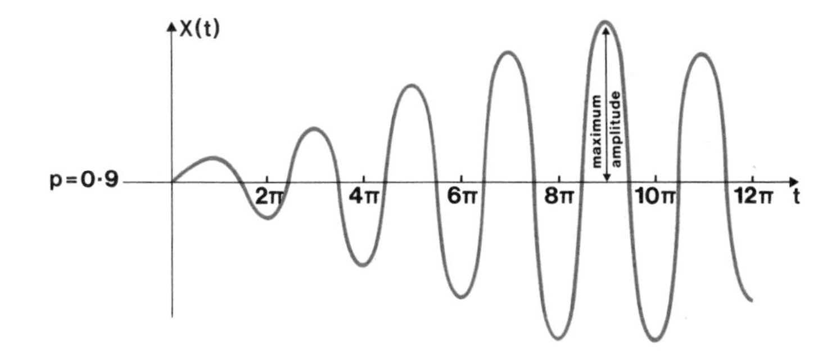

(continued on page 35)

Solution 1

Solution 1

Starting as in the example, we have this time

$$\left(\frac{F_0}{s - mp^2}\right) + \alpha = x_0 \quad \text{and} \quad \beta\omega = v_0,$$

so that

$$\alpha = x_0 - \left(\frac{F_0}{s - mp^2}\right) \quad \text{and} \quad \beta = \frac{v_0}{\omega}.$$

The motion corresponds to the function

$$X: t \longmapsto \left(\frac{F_0}{s - mp^2}\right) \cos pt + \left(x_0 - \frac{F_0}{s - mp^2}\right) \cos \omega t + \frac{v_0}{\omega} \sin \omega t$$

$$(t \in R_0^+) \quad ■$$

Solution 2

Solution 2

In Exercise 31.3.2.2 we found a particular solution of the equation

$$X''(t) + X(t) = \sin pt,$$

namely,

$$X: t \longmapsto \left(\frac{1}{1 - p^2}\right) \sin pt \qquad (t \in R_0^+).$$

Here we have the same equation with $p = \frac{1}{2}$, and so a particular solution is

$$X: t \longmapsto \tfrac{4}{3} \sin \tfrac{1}{2}t \qquad (t \in R_0^+).$$

The general solution of the equation of motion, obtained by adding to this particular solution the general element of the kernel of the operator

$$X \longmapsto X'' + X,$$

is

$$X: t \longmapsto \tfrac{4}{3} \sin \tfrac{1}{2}t + \alpha \cos t + \beta \sin t \qquad (t \in R_0^+).$$

To use the initial conditions we note that

$$X(0) = \tfrac{4}{3} \sin 0 + \alpha \cos 0 + \beta \sin 0 = \alpha$$

and

$$X'(0) = \tfrac{2}{3} \cos 0 - \alpha \sin 0 + \beta \cos 0 = \tfrac{2}{3} + \beta,$$

so that $\alpha = 0$ and $\beta = -\frac{2}{3}$, if the particle starts from rest at the origin. Substituting these values of α and β in the above general solution, we obtained the required solution

$$X: t \longmapsto \tfrac{4}{3} \sin \tfrac{1}{2}t - \tfrac{2}{3} \sin t \qquad (t \in R_0^+).$$

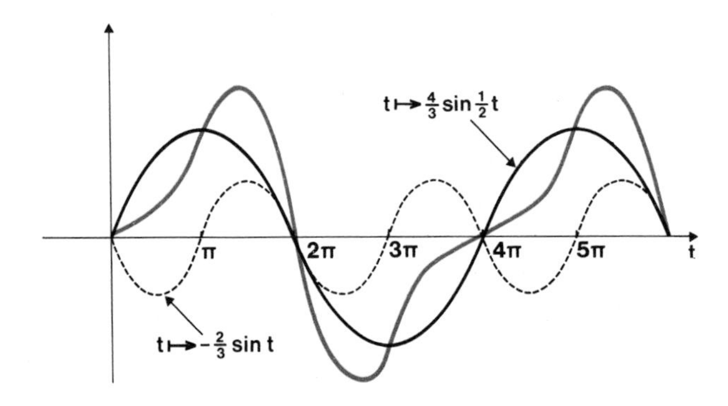

■

Solution 3

Solution 3

(i)

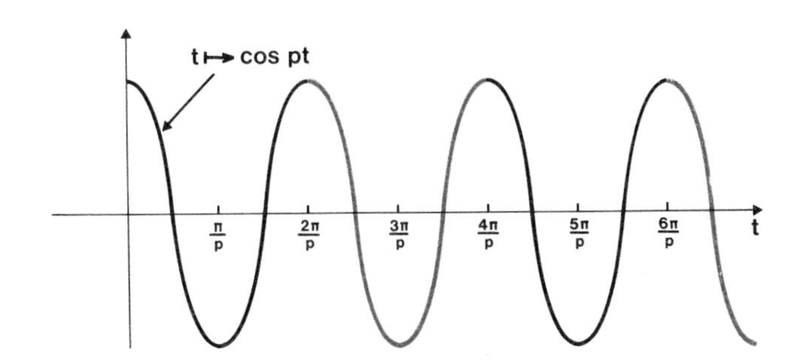

We see that the section of the graph in the interval $\left[0, \frac{2\pi}{p}\right[$ repeats itself again and again and that no smaller interval has this property. The period is therefore $2\pi/p$.

(ii) We have not really drawn the graph far enough in Solution 2 to see the period, but it can be worked out easily.

The $\sin \frac{1}{2}t$ term has period 4π and the $\sin t$ term has period 2π. So the complete function has period 4π. ■

(continued from page 33)

To get a simple estimate of the maximum amplitude, let us assume that this is achieved at a value of t such that one of the two terms $\cos \omega t$ and $\cos pt$ is close to $+1$ and the other is close to -1. Then $|\cos pt - \cos \omega t| \simeq 2$ and so the amplitude is about $\left(\frac{2F_0}{s - mp^2}\right)$. This quantity is plotted against p in the next diagram.

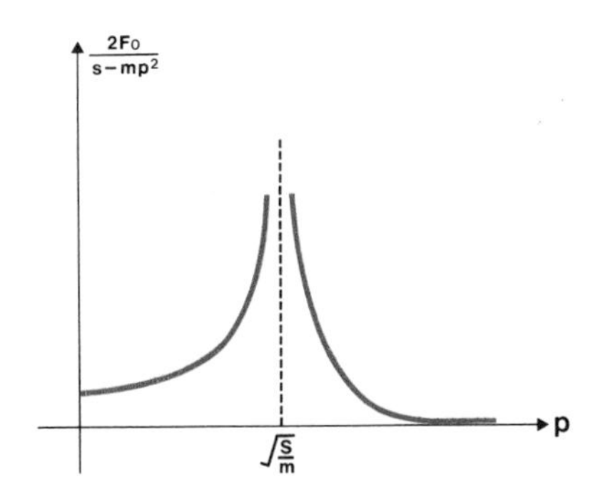

We see that the estimated maximum amplitude can indeed be very large when p is close to ω.

What happens if p is actually equal to ω, so that the applied force has exactly the same period as the free vibrations? The solution we have been using for the equation of motion was based on the assumption that $p^2 \neq s/m$, so it does not cover this case. The equation of motion is now

(taking $F_0 = s = m$ for simplicity)

$$X''(t) + X(t) = \cos t$$

and it is no longer possible to satisfy the equation by taking $X(t) = k \cos t$ with k a number: this form for X makes the left side zero, not $\cos t$, whatever k we use.

There are various methods for finding particular solutions in these exceptional cases. One of them is indicated in the appendix. All we are interested in here is the result, however, which is that the function

$$X : t \longmapsto \tfrac{1}{2}t \sin t \qquad (t \in R_0^+)$$

is a particular solution of the above equation. The graph of this function is shown in the diagram. Notice how it starts very similarly to the graph for $p = 0.9$. The difference is that this one continues to grow for ever: it never reaches a maximum.

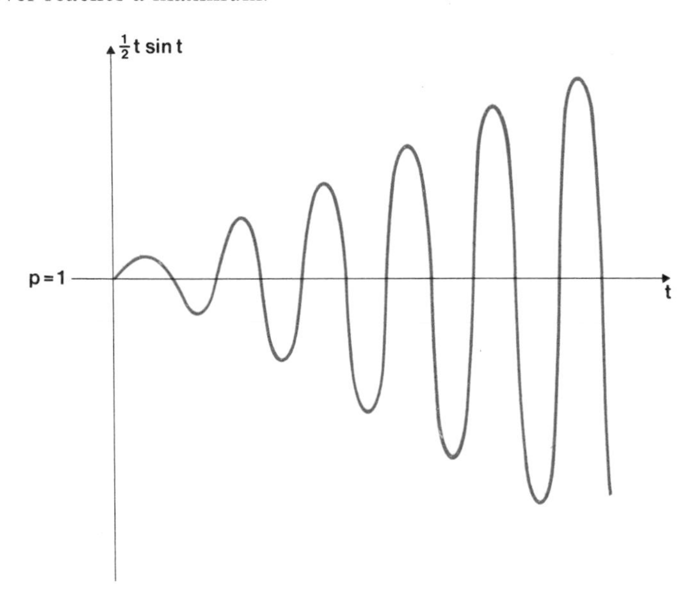

Exercise 4

Exercise 4
(3 minutes)

Verify that

$$X : t \longmapsto \tfrac{1}{2}t \sin t$$

is a particular solution of

$$X''(t) + X(t) = \cos t.$$

■

Exercise 5

Exercise 5
(4 minutes)

Write down the general solution of

$$X''(t) + X(t) = \cos t,$$

and choose the particular solution describing a particle that starts at time 0 with velocity v_0 and displacement x_0. ■

Exercise 6

Exercise 6
(5 minutes)

Obviously it is not *really* possible for vibrations to grow to arbitrarily large amplitude. The prediction of our mathematical model is that they will do so in the case $p = 1$. What features of the physical situation, omitted from our mathematical model, ought to be included if the model is to give a more realistic description of the case $p = 1$? ■

Discussion
* *

In fact, the equation of motion used in this text is a good approximation only for small displacements; for large displacements some more complicated mathematical model would have to be used.

Nevertheless, the model we have been studying does correctly predict and describe the initial growth of the oscillations, before they become too violent.

Postscript

There are some startling real-life examples of resonance — some useful and some disastrous! An example of the former type is an electrical circuit tuned to receive signals of one particular type — such as a radio transmission. The circuit (radio receiver) responds only to the stimulus of a particular frequency — the resonance frequency.

On the other hand, in 1940 the suspension bridge over the Tacoma gorge in North America was destroyed by winds which forced it to vibrate at its natural frequency. There is a famous film taken by a bystander which shows the oscillations building up until the bridge collapses. Some musical instruments illustrate resonance; if a suitable note is played near a violin, the corresponding string will vibrate in sympathy. Slightly different, but in the same context, the famous soprano Dame Nellie Melba was able to shatter a wine glass by singing a sufficiently high note.

Another example is that soldiers must "break step" when marching over a Bailey bridge. If they did not, the force exerted on the bridge might coincide with the natural frequency, and resonance could destroy the bridge.

Love Song of Two Springs

"I love the way you curl",
said the coil to the
spring. The spring
said, "Oscillations
are the very latest
thing." "There's a
period of time, which
is Frequently in Spring,
when external forces
resonate, to get you in
the Swing." "But, in
Autumn," said she
coyly, "When the
friction is
increasing, The
rainfall and
the dampness
have the
amplitude
decreasing."
"So for now
we'll live in
harmony, Our
functions
complementing,
And when oscill-
ations die
away, Then
we can
Rust
In
P
e
a
c
e."

Solution 4

Solution 4

If

$$X(t) = \tfrac{1}{2}t \sin t,$$

then

$$X'(t) = \tfrac{1}{2} \sin t + \tfrac{1}{2}t \cos t,$$

$$X''(t) = \tfrac{1}{2} \cos t + \tfrac{1}{2} \cos t - \tfrac{1}{2}t \sin t,$$

and so

$$X''(t) + X(t) = \cos t \qquad \text{as required.}$$ ■

Solution 5

Solution 5

The general solution is

$$X : t \longmapsto \tfrac{1}{2}t \sin t + \alpha \cos t + \beta \sin t,$$

whence

$$X(0) = \alpha$$

$$X'(0) = \beta.$$

If the particle starts at time 0 with velocity v_0 and displacement x_0, we have $\alpha = x_0$, $\beta = v_0$, and

$$X : t \longmapsto \tfrac{1}{2}t \sin t + x_0 \cos t + v_0 \sin t.$$ ■

Solution 6

Solution 6

The main idealizations made in the model are

(i) friction is neglected;
(ii) it is assumed that, however large the displacements are, the force on the particle is $-sX(t)$ + (impressed force);
(iii) the oscillations never reach the support.

Thus we are assuming that the particle does not hit any other objects (e.g. its support), that the properties of the spring do not change with time (in particular, it doesn't break) and that the force in the spring varies linearly with the displacement $X(t)$, no matter how violently it is stretched and compressed. ■

31.4 APPENDIX (not part of the course)

31.4 Appendix *

Finding a particular solution of $X''(t) + X(t) = \cos t$

For definiteness, we look for the particular solution that corresponds to a motion starting from rest at the origin. We use a limiting process in which p approaches 1, applied to the corresponding solution of the equation

$$X''(t) + X(t) = \cos pt \qquad (t \in R_0^+).$$

As shown in the text, with

$$F_0 = s = m = 1,$$

the solution of this equation with

$$X(0) = X'(0) = 0$$

is

$$X : t \longmapsto \frac{\cos pt - \cos t}{1 - p^2} \qquad (t \in R_0^+).$$

This *suggests* that

$$X : t \longmapsto X(t) = \lim_{p \to 1} \left(\frac{\cos pt - \cos t}{1 - p^2} \right) \qquad (t \in R_0^+)$$

may be a solution of $X''(t) + X(t) = \cos t$. To test this assumption, we evaluate the limit. This can be done using some manipulations based on the identity*

$$\cos(u - v) - \cos(u + v) = 2 \sin u \sin v$$

which, with u and v chosen to make $u + v = t$ and $u - v = pt$, gives

$$\cos pt - \cos t = 2 \sin \left(\frac{1 + p}{2} \right) t \sin \left(\frac{1 - p}{2} \right) t.$$

We find

$$\lim_{p \to 1} \left(\frac{\cos pt - \cos t}{1 - p^2} \right)$$

$$= \lim_{p \to 1} \frac{2 \sin \left(\frac{1 + p}{2} \right) t \sin \left(\frac{1 - p}{2} \right) t}{(1 + p)(1 - p)}$$

$$= \lim_{p \to 1} \left(\frac{2 \sin \left(\frac{1 + p}{2} \right) t}{1 + p} \right) \times \lim_{p \to 1} \left(\frac{\sin \left(\frac{1 - p}{2} \right) t}{\left(\frac{1 - p}{2} \right) t} \right) \times \frac{t}{2}$$

(by the product rule for limits (see *Unit 7*)).

$$= \frac{2 \sin t}{2} \times 1 \times \frac{t}{2} \qquad \left(\text{since } \lim_{x \to 0} \frac{\sin x}{x} = 1, \text{ as shown in Appendix II of } \textit{Unit 12, Differentiation I} \right)$$

$$= \frac{t \sin t}{2}$$

This is the form for $X(t)$ that you are asked to consider in the text.

* This is not the simplest method, but it is the one requiring the least background knowledge.

Unit No.		Title of Text
1		Functions
2		Errors and Accuracy
3		Operations and Morphisms
4		Finite Differences
5	NO TEXT	
6		Inequalities
7		Sequences and Limits I
8		Computing I
9		Integration I
10	NO TEXT	
11		Logic I — Boolean Algebra
12		Differentiation I
13		Integration II
14		Sequences and Limits II
15		Differentiation II
16		Probability and Statistics I
17		Logic II — Proof
18		Probability and Statistics II
19		Relations
20		Computing II
21		Probability and Statistics III
22		Linear Algebra I
23		Linear Algebra II
24		Differential Equations I
25	NO TEXT	
26		Linear Algebra III
27		Complex Numbers I
28		Linear Algebra IV
29		Complex Numbers II
30		Groups I
31		Differential Equations II
32	NO TEXT	
33		Groups II
34		Number Systems
35		Topology
36		Mathematical Structures